2023 NIGHT SKY ALMANAC

A Month-by-Month Guide to North
America's Skies from The Royal
Astronomical Society of Canada

Nicole Mortillaro

FIREFLY BOOKS

A FIREFLY BOOK

Published by Firefly Books Ltd. 2022

First printing

Library of Congress Control Number: 2022934716

Library and Archives Canada Cataloguing in Publication
Title: 2023 night sky almanac : a month-by-month guide to North America's skies
from the Royal Astronomical Society of Canada / Nicole Mortillaro.
Other titles: Two thousand twenty-three night sky almanac | Twenty twenty-three
night sky almanac
Names: Mortillaro, Nicole, 1972- author. | Royal Astronomical Society of Canada,
issuing body.
Description: Includes bibliographical references.
Identifiers: Canadiana 20220199663 | ISBN 9780228103738 (softcover)
Subjects: LCSH: Astronomy—Canada—Observers' manuals. | LCSH: Astronomy—
Canada—Amateurs' manuals. | LCSH: Astronomy—United States—Observers'
manuals. | LCSH: Astronomy—United States—Amateurs' manuals. | LCSH:
Astronomy—Popular works. | LCGFT: Handbooks and manuals.
Classification: LCC QB64 .M6722 2022 | DDC 523—dc23

Published in Canada by
Firefly Books Ltd.
50 Staples Avenue, Unit 1
Richmond Hill, Ontario L4B 0A7

Published in the United States by
Firefly Books (U.S.) Inc.
P.O. Box 1338, Ellicott Station
Buffalo, New York 14205

Project manager/Editor (Firefly Books): Julie Takasaki
Project manager (RASC): Robyn Foret, President of The RASC
Technical editors: David M.F. Chapman FRASC (Emeritus Editor, RASC *Observer's
Handbook*) and James S. Edgar FRASC (Editor, RASC *Observer's Handbook)*
Graphic design: Noor Majeed
Illustrations and sky charts: Peter Kovalik
Printed in China

Canada We acknowledge the financial support
of the Government of Canada.

Contents

Introduction

If you're reading this, you clearly love the night sky and all the joy it brings you.

This guide aims to provide novice and intermediate amateur astronomers with the knowledge they need to enjoy the night sky in 2023, from understanding planets, nebulae, comets, asteroids and meteors, to learning about annual and significant celestial events. It's your pocket guide to the cosmos.

But first, it's important that you understand how we navigate the night sky. Just like for navigation on Earth, astronomers use particular coordinates in the sky to figure out what we're looking at.

The **celestial sphere** is an imaginary sphere with Earth at its center. At any one time, an observer's night-sky view only includes half of this sphere, because the other half is below the horizon.

Earth's axis is tilted at 23.5 degrees to the **plane** of the Solar System — the plane being the orbit of the Earth around the Sun. For observers on the ground, the celestial sphere seems to rotate from east to west; that is why the Sun, for example, rises in the east and sets in the west.

The Earth's axis points less than 1 degree away from **Polaris**, the North Star, in the Northern Hemisphere. It's really important to know where north is, so you can navigate around the sky. Use a star chart, your smartphone or both to find true north at your observing location. (In the south, the Earth's axis points about one

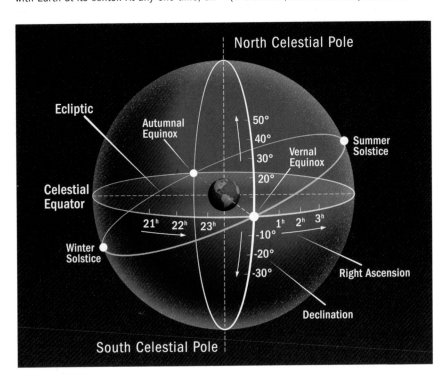

degree away from **Sigma Octantis**, a faint star; given Canada and the United States are in the Northern Hemisphere, however, we will focus on Polaris.) When you look at the sky over long periods of time, Polaris stays still as the other stars rotate around it.

You can easily see this phenomenon by setting up a camera, pointing it toward Polaris, and taking an extended (or long-exposure) photograph of the stars. You will see circular streaks in the sky as other stars rotate around Polaris. Some stars are so close to the North Celestial Pole that they never set. Other stars farther from the pole will rise in the east and set in the west, just like our Sun. Many stars are so far from the pole that they never rise above your horizon.

While beginner astronomers will navigate using a simplified star chart, it's helpful to know some of the terminology we use for finding our way around the sky. That way, as you gain confidence, you can follow professional astronomers in their observations.

The point directly overhead your observing location is called the **zenith**, and the **celestial equator** is directly above the Earth's equator. The point directly below you (under the horizon and opposite to the zenith) is the **nadir**. The line that runs from the north point on the horizon, through the zenith, to the south point on the horizon, is the **meridian**.

To navigate the night sky, astronomers created a type of latitude and longitude called **celestial coordinates**. In astronomy, we use **declination** (Dec.) and **right ascension** (RA). Declination is like latitude on Earth, running from north to south. Right ascension is like longitude on Earth, running from west to east. Declination is measured in degrees, and right ascension is measured in hours, minutes and seconds.

Star trails with Polaris at the center

As Earth rotates, the apparent position of most of the stars changes. More advanced astronomers may want to learn more about exactly where to find faint stars in their telescopes. For these situations, we navigate using **altitude** (angular elevation above the horizon, between 0 degrees at the horizon and 90 degrees at the zenith) and **azimuth** (the number of degrees clockwise from due north). But if you're just starting out, don't worry yet about these advanced observing techniques. A simple star chart will help you find the constellations and the planets.

If you want to look for planets, you should also pay attention to the **ecliptic**. That is the path the Sun takes through the constellations, and the Moon and planets do not stray far from that path.

Handy Sky Measures

Astronomers need to know how far apart things are in the sky. And they do this using angular degrees.

It's not hard to imagine the sky as a sphere that measures 360 degrees — after all, space surrounds Earth on all sides. Standing in one spot on Earth, if you trace the sky from horizon to horizon, that would equal 180 degrees. Remember, the other 180 degrees is under the horizon.

If you want to measure the distances between two objects — say, between the Moon and Venus as they appear together in the sky — you can use your hand as a measuring tool. It all lies within your fingers.

Hold your hand at arm's length. The width of your pinky finger equals 1 degree. The width of your three middle fingers held at arm's length equals five degrees; a closed fist is 10 degrees; the distance between the tip of your index finger and the tip of your pinky is 15 degrees; and the distance between your thumb and pinky is roughly 25 degrees.

This is particularly helpful when trying to see how high or low something is above the horizon.

You can practice measuring the degrees with stars found in the Big Dipper. The chart shows how to find the Big Dipper using Polaris.

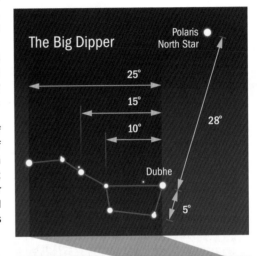

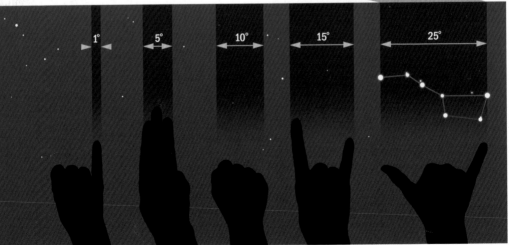

A view of Polaris and the Big Dipper
over Dinosaur Provincial Park, Alberta

Binoculars and Telescopes

Using your eyes to move around the sky allows you to see the Moon, the planets, the Sun and even some objects outside of our Solar System. However, having binoculars or a telescope opens up the Universe to you.

Binoculars are an amazing first tool for seeking out the marvelous wonders of the night sky. A good pair of binoculars can reveal the intricacies of the Moon, showing its peaks and valleys, or uncover the daily motion of Jupiter's moons as they dance around our Solar System's largest planet. A decent telescope can make these objects appear bigger and show details of star clusters or nebulae (gas clouds in space).

But the question that often arises is what type of binoculars or telescope should I get?

For binoculars, it's important to understand two things: the **magnification** (or power) and the **aperture**. Binoculars are usually represented by two numbers, separated by an "×." Two example binocular numbers are 7×50 or 10×50. The magnification is the first number, and it represents the number of times larger something will appear compared to viewing it with the naked eye. The second number is the aperture in millimeters (about 3/64ths of an inch), and it represents the diameter of each lens. Good viewing in part depends on how well your binoculars can gather light. If your binoculars have a larger second number, they have a larger aperture and can gather more light from distant objects. But that comes with a cost. Bigger binoculars are heavier and, therefore, more difficult to hold, which means you will need a tripod to steady your view.

Another number often given for binoculars is the **field of view**, or FOV. This is how wide you will be able to see in degrees (see "Handy Sky Measures," on page 6). In general, the higher the magnification you use, the smaller your field of view will be. At times, this will mean you need to make a choice. Do you want to zoom in on a particular crater on the Moon, for example, or observe a more spread-out chain of mountains? Making these decisions is difficult for astronomers, too, so don't worry if it feels hard — it gets a bit easier to decide what to do with practice.

The top end of handheld binoculars is typically 7×50. Larger than that, it's best to get a tall tripod if you want the best view possible and to share views with others.

We recommend using binoculars for a few months before investing in a telescope. Once you are comfortable with binoculars, luckily, your knowledge will be useful because telescopes use many of the same definitions for observing.

Remember that light-gathering ability is important; a typical beginner's telescope usually ranges between 50 to 150 millimeters (2 to 6 inches) in aperture, the diameter of the main lens or mirror. The bigger the **aperture**, the more light the lens will gather, and the brighter the view will be. Invest in a sturdy tripod, too, so that the telescope will not shake when you use it.

The **magnification** of a telescope depends on the eyepiece used; magnification is calculated by dividing the focal length of the telescope by the focal length of the eyepiece (using like units). For example, a telescope with a focal length of 1,000 millimeters and a 10-millimeter eyepiece will magnify an object 100 times. The same 10-millimeter eyepiece in a 500-millimeter telescope would magnify an object 50 times.

While a bigger aperture and higher magnification may seem like the best way to go, it's best not to get too caught up in choosing a telescope with the highest numbers. What is important is how you plan to use it. Since most people are unlikely to have backyard observatories, the most important thing may be portability. Too heavy a telescope means you're unlikely to haul it out to the backyard or to a vacation spot, far from city lights.

When it comes to telescopes, there is a wide array of choices. Some of the most popular are refractors, reflectors and compound telescopes. Each type has its own benefits, and it all depends on what you prefer. We therefore recommend you try testing out different telescopes at a local star party held by astronomy groups before making the investment. Alternatively, check reputable astronomy magazines or forums for recommendations for beginners; in most cases, all it takes is a little Internet searching and some patience.

A **refractor telescope** has a front lens that focuses light to form an image at the back, and an eyepiece that acts like a magnifying glass to allow your eye to focus on that image. Refractors tend to be more reliable, as their lenses are fixed in place and, therefore, don't get out of alignment as easily as some other types of telescopes. Refractors are best used for planetary and lunar observing as well as viewing double stars.

A **reflector telescope** uses a mirror to gather and focus light. The advantage for reflectors is they don't usually suffer from **chromatic aberrations**, when light of different wavelengths (i.e., colors) doesn't focus on the same point; this makes them ideal to observe distant objects like star clusters. Chromatic aberrations cause the different colors of the spectrum to split and the image to appear blurry, which isn't great when looking at a group of stars. A **Dobsonian telescope**, which is a variant of a reflector with a simple mount, gives you a far bigger aperture for far less the cost of other telescopes.

Compound or **catadioptric telescopes** combine lenses and mirrors to form an image. The **Schmidt-Cassegrain**, a type of compound telescope, has a compact design that makes it quite popular. With this instrument, an astronomer can get a bigger aperture in a smaller sized telescope. This telescope type is also portable, making it easier to move the equipment into remote areas.

Star parties are great opportunities to check out different telescopes and chat with fellow enthusiasts

Stars

Stars come in many different varieties. Our Sun is a **yellow dwarf star**, on the **main sequence**, meaning that it's converting hydrogen into helium at its core, like most other stars. When a star does this conversion, it releases a tremendous amount of energy. This energy, in the form of sunlight, allows life to thrive on Earth.

The Sun, at 4.5 billion years old, is considered middle-aged. When it dies, five billion years from now, it will at first swell, becoming a **red giant** and engulfing the inner planets, then it will slough off its outer layers to become a **white dwarf**.

One of the most interesting types of stars, some might argue, are **red supergiants**. These colossal stars — roughly 1,400 times the mass of the Sun — have relatively short lifespans.

And when supergiants do die, they do so in a spectacular fashion, in an explosion called a **supernova**. A supernova occurs when a star can no longer convert hydrogen into helium; eventually the core is converted into iron. During a supernova, the star first collapses and then explodes outward, creating even heavier elements.

Betelgeuse, a star found in the left shoulder of Orion, is a red supergiant. While many far-away and faint supernovae have been witnessed in modern history in other galaxies, none have occurred in our own galaxy — the **Milky Way**. When Betelgeuse goes supernova, its brightness will rival the full Moon in our sky. If you're hoping to see Betelgeuse explode, you're not alone. However, estimates peg Betelgeuse's death for some time within the next 100,000 years.

The most common star in the Universe is a **red dwarf**, which is a cool star that's much smaller than the Sun. Astronomers often search for potentially habitable planets around these stars, because the inherent dimness and smaller size of red dwarfs makes it easier to spot planets. The stars, however, can be quite volatile, occasionally releasing a tremendous amount of radiation. Intense radiation is not a friendly process for most life forms.

The **Hertzsprung-Russell diagram** (H-R diagram) was developed in the early 1900s by Ejnar Hertzsprung and Henry Norris Russell, following the research findings by two Harvard computers, Annie Jump Cannon and Antonia Maury. The diagram plots the temperature of stars against their luminosity. Just as we do, stars go through certain stages in their lives.

Top 10 Brightest Stars in the Night Sky	
1.	Sirius
2.	Canopus
3.	Alpha Centauri
4.	Arcturus
5.	Vega
6.	Capella
7.	Rigel
8.	Procyon
9.	Achernar
10.	Betelgeuse

The Harvard Observatory Computers

In 1881, Edward Charles Pickering, the director of the Harvard Observatory, hired a team of women to compute and catalog photographs of the night sky. Their work was incredibly important in providing the foundations of astronomical theory. Annie Jump Cannon, one of the computers, added to work done by fellow computer Antonia Maury and developed a system of classifying stars that is still used today.

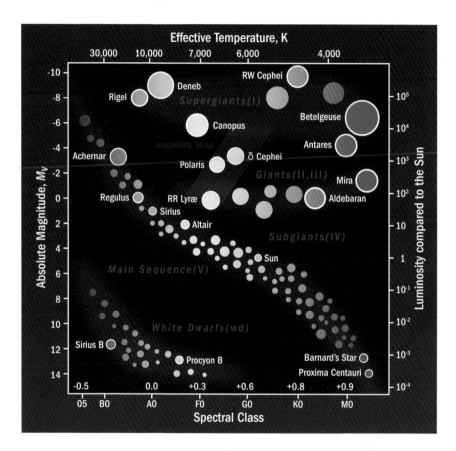

The H-R diagram provides astronomers with the information about a star's current age. Main-sequence stars that are fusing hydrogen into helium — such as our Sun — lie on the diagonal branch of the diagram.

Finally, there's a star's **magnitude**, or apparent brightness. Magnitude is measured on a scale where the higher the number, the fainter it is — and negative numbers are brighter than positive numbers. For example, Sirius, the brightest star in the night sky, measures –1.4 on this scale. Polaris is 2.0, and the Sun is –27. We also use magnitude to measure the brightness of other celestial objects, like the

Moon, planets, asteroids and comets.

Note there is a difference between apparent and absolute magnitude. Apparent magnitude is the brightness of an object that we observe from Earth, but absolute magnitude is the brightness of an object if it were placed 32.6 light-years from Earth. This second measure helps astronomers directly compare the luminosity of objects and is what is used for the H-R diagram. On this scale, Sirius has a magnitude of 1.4, Polaris is –3.6 and the Sun is 4.8.

In astronomy, Greek letters of the alphabet are used to identify stars within a constellation usually from brighter to dimmer.

Constellations

Constellations, a group of stars that make an imaginary image in the night sky, have been around since ancient times. Typically, the images astronomers use are based on Greek, Roman and Arabic mythologies. The International Astronomical Union (IAU) recognizes a total of 88 official constellations.

Some of the most recognizable constellations in the Northern Hemisphere are Orion (the Hunter), Cygnus (the Swan), Leo (the Lion), Gemini (the Twins), Scorpius (the Scorpion) and Ursa Major (the Great Bear).

More recently, there has been more effort to acknowledge constellations of Indigenous peoples. The naming of Indigenous constellations varies around the world, making these constellations regionally distinct. Some groups see

Ursa Major as a bear, or a caribou. The Cree see Corona Borealis as seven birds, and Cepheus as a turtle. To the Navajo, Polaris is Nahookos Bikq, meaning central fire. For some Indigenous peoples, such as the Inuit, the Northern Lights are dancing spirits.

Aside from the constellations, there are also **asterisms**, a group of stars within a constellation (or sometimes from several different constellations) that forms its own distinct pattern. The Big Dipper is probably the most famous asterism, as its stars lie within the constellation of Ursa Major. There's also the Summer Triangle, with bright stars from Cygnus, Lyra (the Lyre) and Aquila (the Eagle), and the Winter Triangle with stars from Orion, Canis Major (the Great Dog) and Canis Minor (the Little Dog).

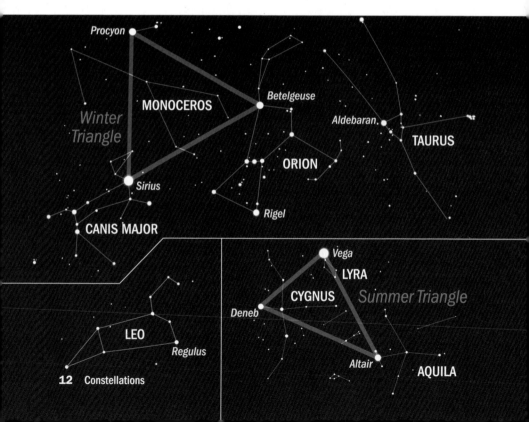

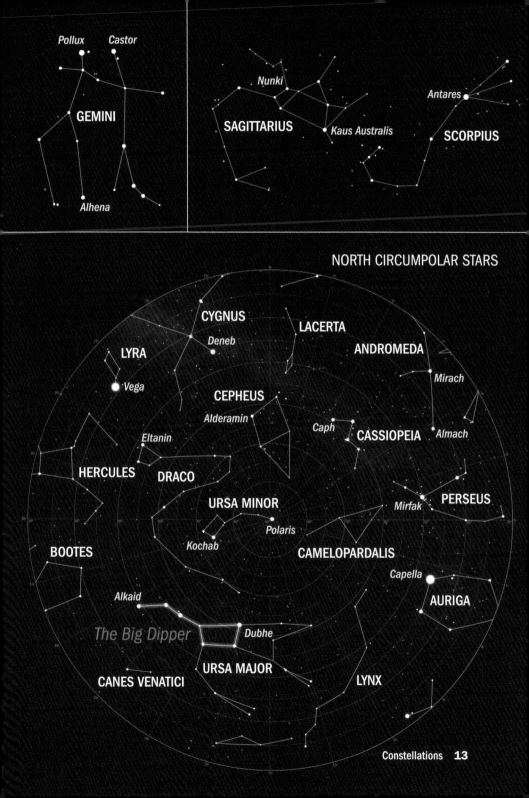

Pollux · Castor

GEMINI

Alhena

Nunki

SAGITTARIUS

Kaus Australis

Antares

SCORPIUS

NORTH CIRCUMPOLAR STARS

CYGNUS
Deneb

LACERTA

ANDROMEDA
Mirach

LYRA
Vega

CEPHEUS
Alderamin

Caph

CASSIOPEIA
Almach

Eltanin

HERCULES

DRACO

URSA MINOR

Mirfak

PERSEUS

BOOTES

Kochab

Polaris

CAMELOPARDALIS

Capella

AURIGA

Alkaid

The Big Dipper

Dubhe

CANES VENATICI

URSA MAJOR

LYNX

Constellations **13**

Comets and Meteors

Comets are icy balls of debris — specifically, dust and ice — left over from the formation of our Solar System. They are sometimes referred to as "dirty snowballs" and can be stunning objects to see in the night sky when tails of dust and ionized gas fan out behind their cores. While we know of many comets, predicting the appearance of a bright one is all but impossible. Comets tend to fall apart before becoming too bright. The Sun's gravity and heat are strong, and comets themselves are very delicate visitors from the outer Solar System or beyond. Many comets literally crumble under the pressure as they dive in toward the Sun and the inner Solar System, where our planet resides.

Bright comets can be marvelous. On August 17, 2014, Comet Lovejoy C/2014 Q2 was spotted as it came toward the inner Solar System. This comet produced a spectacular tail as it neared the Sun. Tails occur when ice **sublimates**, turning directly from a solid into a gas.

There have been other wonderful naked-eye comets that have graced our night sky, including Comet Hyakutake C/1996 B2, Hale-Bopp C/1995 O1, or Comet NEOWISE C/2020 F3. Each comet has the year of its discovery in its official name, so NEOWISE was found in 2020, Hyakutake in 1996, and so on. Many Northern Hemisphere observers were treated to a special show when Comet NEOWISE appeared in the skies in 2020. It was one of the brightest comets to appear in a generation — even city dwellers could spot it through light pollution.

Periodic comets, or ones that we can predict, are given the designation "P," while those that appear unexpectedly are given the designation "C." For example, Halley's Comet, or 1P/Halley, appears roughly every 76 years — a clear P. The last time it passed was 1986; the next time will be in 2061. In general, C comets are brighter than P comets because P comets have sublimated their material into space from repeated trips by the Sun.

A **meteor** (from the Greek *meteoros*, meaning "high in the air") is the light, heat and (occasionally) sound phenomena produced when a

Comet NEOWISE C/2020 F3

The Geminid meteor shower

meteoroid, or small debris left in space from comets or **asteroids** (space rocks), collides with molecules in Earth's upper atmosphere. We also call meteors **shooting stars**.

When a meteoroid enters Earth's atmosphere, the surface of the object is heated. Then, at a height typically between 119.8 and 79.9 kilometers (74.5 miles and 49.7 miles), the meteoroid begins to **ablate**, or lose mass. Meteoroid ablation usually happens through vaporization, although some melting and breaking apart can also occur.

Meteoroids are usually the size of a small pebble, but they range in size considerably; some can be as small as the size of the tip of a ballpoint pen, while more unusually, they can be several meters or kilometers across. If a meteoroid is large enough that it doesn't burn up entirely and reaches the ground, it is called a **meteorite** — and we have recorded instances of meteorites causing damage on Earth.

NASA keeps a sharp eye out for threatening objects in space and, so far, has found nothing of concern. We know a big meteorite strike could happen eventually, though. Just ask the dinosaurs, which were likely wiped out by a meteorite at least 11 kilometers (7 miles) across about 66 million years ago. Thankfully, such events are quite rare, happening every few million years on average.

Meteoroids can be divided into two groups: **stream** and **sporadic meteoroids**. Stream meteoroids have orbits around the Sun that often can be linked to a parent object — in most cases a comet, but it can sometimes be an asteroid. When Earth travels within the orbit of a stream, we get a **meteor shower**. Because all the objects (meteoroids) move in the same direction, they all seem to come from one point in the sky, called the **radiant**. The constellation containing the radiant (or in some cases a nearby star) is what gives a meteor its name, such as the Perseids or Geminids (two of the most active meteor showers).

Sporadic meteors just occur in random parts of the sky, with no radiant.

Meteor Showers in 2023

There is nothing more wonderful than stepping out and looking up at the night sky, only to see a brief streak of light flash against the stars.

Almost monthly, we get major meteor showers. Some showers are best seen from the Northern Hemisphere, while some are better visible in the south, depending on the radiant.

When astronomers talk about the "peak" of a meteor shower, they're referring to the **Zenithal Hourly Rate**, or the ZHR. This is the rate of meteors a shower would produce per hour under clear, dark skies and with the radiant at the zenith. In practice, a single observer will likely see considerably fewer meteors than the ZHR suggests, owing to the presence of moonlight, light pollution, the location of the radiant, poor night vision and inattentiveness.

Here is a list of major meteor showers that you can enjoy simply by staying warm and looking up. No special equipment is needed beyond your eyes; just make sure to give yourself about 20 minutes to adjust to the darkness before searching for meteors.

Quadrantids:
December 27, 2022–January 10, 2023

The Quadrantids might be one of the best meteor showers of the year, with a ZHR of 120 for a brief interval of time. The only thing holding it back from earning the title is that January tends to be cloudy over North America, and the shower's peak has a brief window of six hours, making the average hourly rate closer to 25. But there's good news: the meteors often produce bright **fireballs**, or bright meteors with a magnitude higher than –4.0. This shower's radiant lies between Boötes and Draco. The only downside is that the Moon will be almost full on the peak night.

Parent Object: Asteroid 2003 EH1
2023 Peak Night: January 3–4

Lyrids: April 16–30, 2023

After a dearth of meteor showers in the months of February and March, April brings us the Lyrids. This shower produces a ZHR of 18, so it's not a particularly strong shower, but the meteors that do appear tend to produce fireballs. The Moon won't interfere with observations on the peak night, as it will be a waxing crescent and will set early in the night.

Parent Object: Comet C/1861 G1 (Thatcher)
2023 Peak Night: April 22–23

Eta Aquariids: April 19–May 28, 2023

The Eta Aquariids aren't a stellar show for the Northern Hemisphere, since the radiant rises in early dawn, but they can still produce a ZHR of 20 meteors, if you care to get up early enough. Rather than fireballs, this shower tends to produce long trains through the sky. Unfortunately, a full Moon will interfere with all but the brightest meteors.

Parent Object: Comet 1P/Halley
2023 Peak Night: May 6–7

Perseids: July 17–August 26, 2023

For those in the Northern Hemisphere, the Perseids are considered the best show of the year. The weather is warm and there are fewer clouds at this time of year. The shower's ZHR is 100, though rates of 50 to 75 are more commonly seen on the peak night. The great news is that the Moon will not interfere with the show this year.

Parent Object: Comet 109P/Swift-Tuttle
2023 Peak Night: August 12–13

Orionids: October 2–November 7, 2023

This shower is considered medium strength, though it can sometimes surprise us with more activity. On average, the Orionids produce a ZHR of 10 to 20 meteors, though from 2006 to 2009, the shower produced roughly 50 to 75 an hour. The meteors that enter our atmosphere are fast — roughly 66 kilometers (41 miles) per second — and, as a result, produce long, glowing trains. Fireballs are also possible. The Moon will be illuminated roughly 37 percent, but will set early in the night.

Parent Object: Comet 1P/Halley
2023 Peak Night: October 21–22

Southern Taurids:
September 10–November 20, 2023
The Southern Taurids last for two months and have several minor peaks. Though this shower produces a ZHR of only 5, it can produce some fireballs. The shower is stronger in the Southern Hemisphere, though we can also catch a few meteors in the north. The Moon will not interfere with this shower on the peak night.

Parent Object: Comet 2P/Encke
2023 Peak Night: November 5–6

Northern Taurids:
October 20–December 10, 2023
Like the Southern Taurids, the Northern Taurids last roughly two months and have a ZHR of 5. There are often reports of an increase of fireballs during the period when the two showers overlap. Once again, the sky will be Moon-free for this shower.

Parent Object: Comet 2P/Encke
2023 Peak Night: November 12–13

Leonids: November 6–30, 2023
While the Leonids don't produce a high number of meteors an hour (they have a ZHR of 15), they can produce outbursts, or meteor storms; the most recent such outbursts occurred in 1999 and 2001. Unfortunately, the next outburst isn't expected until 2099. Still, the Leonids do put on a show, with bright meteors and long-lasting trains. A waning Moon rises after midnight local time on the peak night.

Parent Object: Comet 55P/Tempel-Tuttle
2023 Peak Night: November 17–18

Geminids: December 4–17, 2023
Due to the weather — with increasing cloudiness and chillier temperatures — December isn't an ideal time to catch a meteor shower. But December is the month with the most active meteor shower of the year, called the Geminids. This shower has a ZHR of 150, and the meteors it produces are usually bright and colorful. There's great news for this shower: The Moon will not be up, so you'll be able to see faint meteors.

Parent Object: Asteroid 3200 Phaethon
2023 Peak Night: December 14–15

Ursids: December 17–26, 2023
Right on the heels of the Geminids is the often-forgotten Ursid meteor shower. The ZHR for this shower is 10, but sometimes you might catch an outburst that could produce a ZHR of 25. Unfortunately, an almost-full Moon will interfere with all but the brightest meteors.

Parent Object: Comet 8P/Tuttle
2023 Peak Night: December 22–23

The Moon

The Moon is most often the first astronomical object to capture anyone's attention in the night sky. It's also likely the first object most people see up close, whether it be through binoculars or a telescope.

Our nearest celestial neighbor has been an object of fascination since humans first looked up to the sky. Early scientists in ancient observatories carefully tracked the cycles of the Moon. Some 23 centuries ago, Aristarchus of Samos carefully observed a total lunar eclipse and derived impressively accurate measurements of the Moon's diameter. He estimated its diameter was one-third that of Earth. He was close: the actual percentage is 27.2, and its diameter is roughly 3,540 kilometers (2,200 miles).

It takes about 29.5 days for the Moon to go through its cycle of **phases**. When you can't see the Moon at all, it's called a **new Moon**.

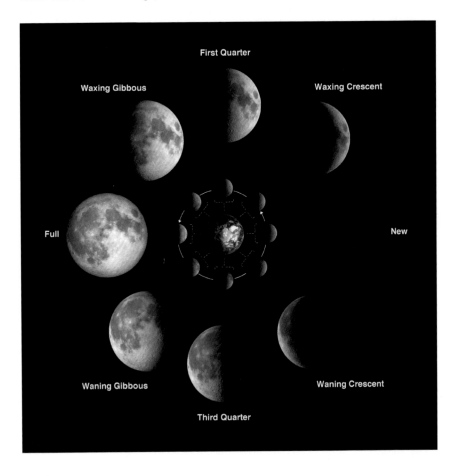

When the Moon is fully illuminated, it is a **full Moon**. The list of phases is new Moon, waxing crescent Moon (when the Moon is just a thin crescent), first quarter Moon (when it appears half full), waxing gibbous Moon (when the Moon is between half-full and full), then full Moon — before the process runs in reverse from full to new: waning gibbous Moon, third or last quarter Moon, waning crescent Moon, new Moon.

The Moon is tidally locked with Earth, meaning that we only ever get to see one face of it. The Sun does shine on the other side of the Moon when we can't see it, so don't call the far side of the Moon the dark side.

While it may seem that we always see the same Moon features month after month, that's not entirely true. The Moon oscillates from our perspective, a process called **lunar libration**. As a result, every so often we see some small part of the Moon that we don't always see.

Observing the Moon

You don't need a telescope to enjoy the Moon. If observed even briefly on a regular basis, Earth's only natural satellite has much to offer the naked-eye observer. Keep an eye on it, and you will notice things like its wandering path through the constellations, the changing phases, frequent **conjunctions** (where two objects appear close together in the sky) with planets or bright stars, occasional eclipses, lunar libration, earthshine (illumination on the Moon reflected from our planet) and other wonderful atmospheric effects.

If you happen to take a look through a telescope, binoculars or even a camera, our closest astronomical target offers an amazing amount of interesting detail. There are countless features on the lunar nearside, and more than 1,000 have been formally named by the IAU. Many of the greatest names in astronomy, exploration and discovery are commemorated through these names. For extra dramatic effect, try looking at a lunar feature when the **terminator** — the line between darkness and sunlight — runs nearby the feature. The extra shadow could make it easier to see details in the object you're interested in looking at.

An estimated three to four billion years ago, our Solar System was bombarded by debris for hundreds of millions of years. This period is called the Late Heavy Bombardment, and you can see many of the scars from that time left over on the Moon.

Because the Moon has almost no atmosphere and no tectonic activity, those scars have remained over billions of years. Some of that activity created deep **impact basins** that filled with lava flows and hardened into basaltic rock. These largely circular regions were termed **maria**, or seas, by early astronomers because they resembled Earth's oceans.

The brighter white areas, which are highlands that surround the maria and dominate the southern portion of the Earth-facing hemisphere, feature many ancient **impact craters**.

After the bombardment slowed to a trickle, the Moon remained relatively free from intense impact activity — meaning the Moon has changed very little since billions of years ago. But, just like on Earth, meteorites still periodically slam into the Moon today.

The Moon map on the next two pages shows a few interesting features for you to target as you observe the Moon.

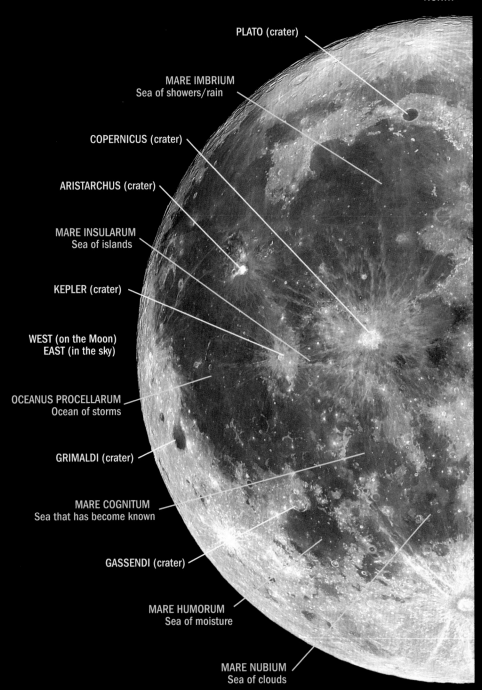

PLATO (crater)

MARE IMBRIUM
Sea of showers/rain

COPERNICUS (crater)

ARISTARCHUS (crater)

MARE INSULARUM
Sea of islands

KEPLER (crater)

WEST (on the Moon)
EAST (in the sky)

OCEANUS PROCELLARUM
Ocean of storms

GRIMALDI (crater)

MARE COGNITUM
Sea that has become known

GASSENDI (crater)

MARE HUMORUM
Sea of moisture

MARE NUBIUM
Sea of clouds

SOUTH

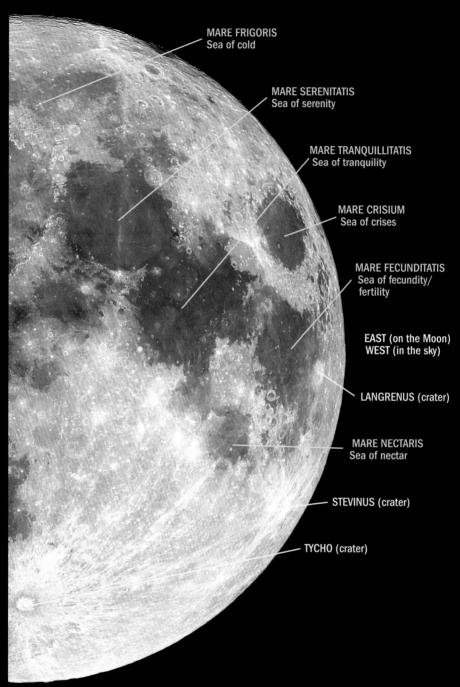

NORTH

MARE FRIGORIS
Sea of cold

MARE SERENITATIS
Sea of serenity

MARE TRANQUILLITATIS
Sea of tranquility

MARE CRISIUM
Sea of crises

MARE FECUNDITATIS
Sea of fecundity/
fertility

EAST (on the Moon)
WEST (in the sky)

LANGRENUS (crater)

MARE NECTARIS
Sea of nectar

STEVINUS (crater)

TYCHO (crater)

SOUTH

The Moon **21**

The Sun

Our Sun highly influences Earth. It is connected with our seasons, weather, ocean currents and climate, and its power makes life itself possible.

Unlike Earth, the Sun isn't a solid body, and due to that, different parts of it rotate at different rates. At the equator, for example, it spins roughly once every 25 days, while the polar regions take about 30 days to complete a rotation.

In the core of the Sun, where hydrogen atoms fuse to make helium, temperatures are a searing 15 million degrees Celsius (27 million degrees Fahrenheit). The tremendous amount of energy generated at the core — **thermonuclear fusion** — is carried outward by radiation, taking roughly

Take Care of Your Eyesight

Do not observe the Sun without special protective equipment, such as a certified solar filter that covers your eyes or telescope aperture entirely. Unprotected observations even for a few seconds could damage your eyesight permanently.

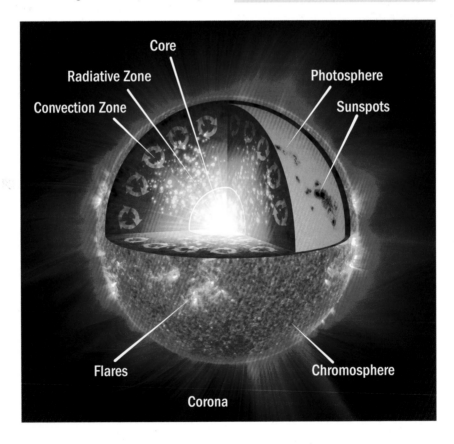

Core

Radiative Zone

Convection Zone

Photosphere

Sunspots

Flares

Chromosphere

Corona

170,000 years to get from the core to the top of the convective/convection zone. As hot as the core is, at the surface, it's a much different story — the temperature is a much "cooler" 5,500 degrees Celsius (10,000 degrees Fahrenheit).

The Sun also has a roughly 11-year cycle, that has a **maximum** (a period of increased solar activity) and a **minimum** (a period of low solar activity). During a maximum, the number of **sunspots**, which are cooler regions on the surface of the Sun, increases. Sometimes, the magnetic field lines of these sunspots can become entangled, finally snapping and releasing a tremendous amount of radiation into space. This process is called a **solar flare**. And often a **coronal mass ejection (CME)** can follow a flare, with charged particles of the Sun speeding outward into space. If these particles reach Earth, they can disrupt radio transmissions, damage satellites and, more positively,

A New Solar Cycle

At the end of 2020, we started Solar Cycle 25. After a quiet solar minimum, we can now expect to see more solar activity. Solar Cycle 24 was the weakest cycle in the last 200 years.

interact with our magnetic field. When the particles do interact with the field, they can produce the beautiful Northern Lights or aurora borealis.

Though beautiful, it must be said that these outbursts from the Sun can also cause power outages, as was experienced in Quebec in 1989. As such, astronomers are keen to better understand our nearest star using many spacecraft — such as the Parker Solar Probe, which was launched in 2018. Keep watching the eight-year adventure of the Parker Solar Probe as it studies the Sun's activity.

The Northern Lights

Observing the Sun

The Sun, our closest star, is a wonderful object to observe with any type of telescope, if used safely. The only safe filter is the kind that covers the full aperture of the telescope, at the front, allowing only 1/100,000th of the sunlight through the telescope. Without this filter, you risk permanently damaging your telescope or, far worse, your eyes. Ensure that the filter material is specifically certified as suitable for solar observing, and *take great caution every time you observe the Sun.*

The preferred solar filters are made of Thousand Oaks glass or Baader film. Thousand Oaks glass gives the Sun a golden-orange color, while Baader film gives the Sun a more natural white look, which can provide good contrast if there are any bright spots, or **plages**, on the Sun. Other solar-filter options include lightweight Mylar film (which gives the Sun a blue-white tint), and eyepiece projection. Projecting the image from the eyepiece onto white cardstock paper, or a wall, produces an image that can be safely shared with others.

While it might seem like the Sun is just this boring yellow ball in the sky, there are many things to see on its surface, including **prominences**, **filaments**, flares and sunspots. These phenomena result from the strong magnetism within the Sun, which erupts to the surface.

Prominences are solar plasma ejections, some of which fall back to the Sun in the form of teardrops or loops; against the Sun's disk, prominences viewed from above appear as dark filaments. Sunspots are cooler regions of the Sun that appear dark on the face of the surrounding surface. If the highly magnetic lines of sunspots intertwine, they can snap and cause a solar flare — a bright, sudden eruption of energy that can last from a few minutes to several hours. It should be noted that to see prominences and some other solar features, you need very specialized filters.

If you do not have the proper equipment to observe the Sun, you can always visit the Solar and Heliospheric Observatory (SOHO) and the Solar Dynamics Observatory (SDO) websites (see the list of resources on page 120) to see daily images of the Sun.

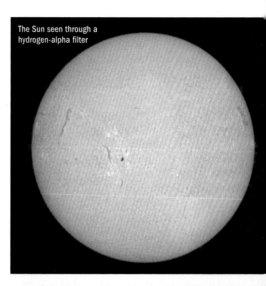

The Sun seen through a hydrogen-alpha filter

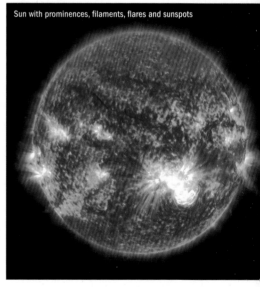

Sun with prominences, filaments, flares and sunspots

Eclipses

Eclipses — whether they're solar or lunar — are seemingly magical occurrences. In fact, that's exactly what many ancient civilizations believed. Several legends suggested a celestial being devoured the Sun during a solar eclipse. For the Chinese, that being was either a dog or a dragon. The Vikings believed it was two wolves called Hati and Skoll. For the Vietnamese, it was a giant frog. Today, we understand that eclipses are awe-inspiring celestial events resulting from the movement and positions of the Earth, Moon and Sun. There can be as many as seven combined eclipses (i.e. solar and lunar) in a year, but no fewer than four.

A solar eclipse is truly a chance occurrence. The Sun's diameter is roughly 400 times that of the Moon, but the Sun is also about 400 times farther away from Earth than the Moon is from Earth. This means the Moon and the Sun appear roughly the same size in the sky, with only about 1/10th of a degree across of difference. In a **total solar eclipse**, the new Moon covers the entire face of the Sun, revealing the Sun's stunning corona and prominences. As totality begins and ends, sunlight peeking around the mountainous limb of the Moon creates fleeting effects such as the Diamond Ring and Baily's Beads. If the Moon were farther away or smaller, we wouldn't get this marvelous sight. In fact, the elliptical nature of both the Earth's orbit around the Sun and the Moon's orbit around the Earth means that the apparent diameters

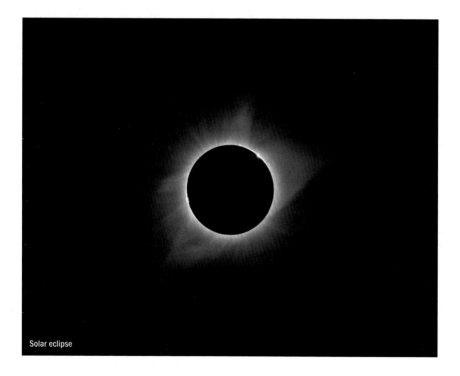

Solar eclipse

of both objects vary significantly. Consequently, it is common for eclipse durations to vary by several minutes.

It should be noted that the totality only occurs in a very narrow path. Outside of this path, observers will only see a partial eclipse. That's why solar eclipse enthusiasts will travel great distances to take in the captivating sight of a totality. Remember, do not look at a solar eclipse unless you use special equipment recommended by The Royal Astronomical Society of Canada or the American Astronomical Society. The only time it is safe to look at a solar eclipse with the naked eye is when it is at its total phase, with the Sun completely covered by the Moon.

Because the Moon orbits at a 5-degree inclination to the ecliptic, we don't get a solar eclipse every month. Every year, there are two 36-day eclipse seasons during which either solar or lunar eclipses can occur, and the eclipse seasons move backward in time every year by 19 days. Also, because the Moon's orbit is elliptical, the angular size of the Moon varies along its path, and sometimes we have an **annular solar eclipse** (or "ring of fire"), when the Moon's disk does not completely cover the Sun's disk. There may be anywhere between zero to two total or annular solar eclipses in a given year. There may also be **partial solar eclipses**, when the Moon's central shadow entirely misses the Earth.

As for lunar eclipses, these occur when Earth is situated directly between the Sun and the full Moon. The Moon drifts through Earth's two shadows: the **penumbra**, which is the fainter, outer shadow, and the deeper shadow that is called the **umbra**. A **penumbral eclipse** is difficult to see with the naked eye as the brightness of the Moon doesn't appear to dim much. But a **partial** or **total lunar eclipse**, when the Moon passes through the umbra, is much more dramatic. During a total lunar eclipse, the Moon can turn an orange-red color, depending on Earth's atmosphere. There may be anywhere between zero to three total lunar eclipses in a given year. Typically, they occur about two weeks before or after a solar eclipse.

While not all eclipses will be visible from North America, you can watch online on sites like SLOOH or The Virtual Telescope Project. You can also see astronomer Fred Espenak's webpage eclipsewise.com for a comprehensive treatment of solar and lunar eclipses.

Eclipses in 2023

This year we have a few eclipses, including a rare hybrid solar eclipse. During a hybrid solar eclipse, a viewer may either see it as a total solar eclipse or an annular solar eclipse, depending on where along the eclipse path they are looking from. Unfortunately this year's hybrid solar eclipse will not be visible from North America.

In general, the 2023 eclipse year will be rather uneventful for North Americans, but we will be rewarded for our patience by another "Great North American Solar Eclipse" on April 8, 2024, when the line of totality crosses northern

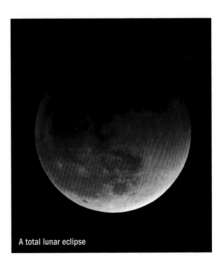

A total lunar eclipse

Mexico, the central U.S. and eastern Canada.

All the times listed are in UTC, or Coordinated Universal Time. See page 40 to learn how to calculate your local time from the time shown in UTC. Also, though we have noted when the Moon enters the penumbra (as this is a part of the stages of a lunar eclipse), it is not noticeable to the human eye. For precise details and extra resources on eclipses, visit www.eclipsewise.com.

Lunar Eclipses

Penumbral Lunar Eclipse: May 5, 2023

Penumbral eclipses occur when the Moon passes through Earth's outer shadow, called the penumbra. Unfortunately, you likely won't notice much of a difference in the Moon's brightness, as these eclipses are less dramatic than either total or partial lunar eclipses. This eclipse will be seen across southern and eastern Europe, Africa and Asia as well as Australia and Indonesia (Oceania).

Eclipse times (UTC)	
Moon enters the penumbra (P1)	15:15
Greatest eclipse	17:24
Moon exits penumbra	19:33

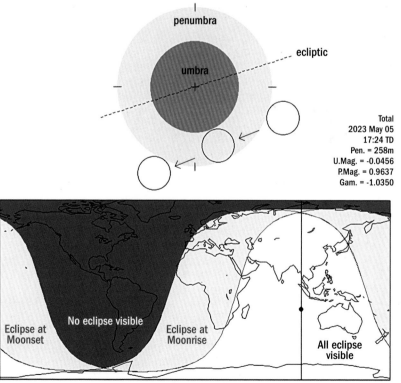

penumbra

ecliptic

umbra

Total
2023 May 05
17:24 TD
Pen. = 258m
U.Mag. = -0.0456
P.Mag. = 0.9637
Gam. = -1.0350

Eclipse at Moonset

No eclipse visible

Eclipse at Moonrise

All eclipse visible

Thousand Year Canon of Lunar Eclipses © 2014 by Fred Espenak

Partial Lunar Eclipse: October 28, 2023

While this is considered a partial lunar eclipse, only a small fraction of the Moon will enter the darkest part of Earth's shadow, the umbra. Most of this eclipse will occur as a penumbral lunar eclipse. This is to say that it will be difficult to notice much of a change in the color or brightness of the Moon. The eclipse will be visible across the eastern Americas, Africa, Europe, Asia and Oceania.

Eclipse times (UTC)	
Moon enters the penumbra (P1)	18:03
Partial eclipse begins (U1)	19:36
Greatest eclipse	20:15
Partial eclipse ends (U4)	20:53
Moon exits penumbra	22:27

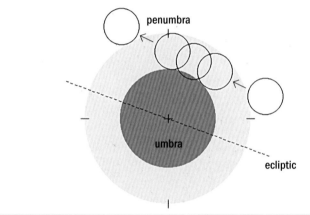

Partial
2023 Oct 28
20:15 TD
Par. = 77m
Gam. = 0.9472
U.Mag. = 0.1221
P.Mag. = 1.1181

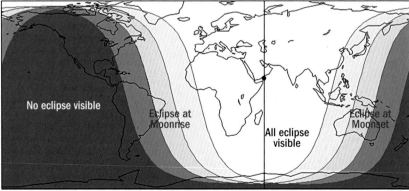

Thousand Year Canon of Lunar Eclipses © 2014 by Fred Espenak

Solar Eclipses

Hybrid Solar Eclipse: April 20, 2023

This eclipse will be visible beginning from the Indian Ocean, across Indonesia, Papua New Guinea and Australia, to the Pacific Ocean. It will begin as an annular solar eclipse, change to a total solar eclipse and then change back to an annular eclipse before the path ends. It will be seen as a partial solar eclipse in Southeast Asia, Malaysia, Indonesia, Australia, Philippines and New Zealand. The eclipse begins at 1:35 UTC and ends at 6:59 UTC.

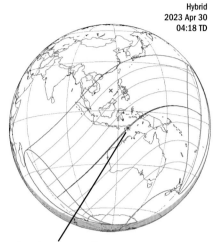

Hybrid
2023 Apr 30
04:18 TD

point of greatest eclipse

Thousand Year Canon of Solar Eclipses
© 2014 by Fred Espenak

Annular Solar Eclipse: October 14, 2023

This annular solar eclipse will be visible across the western U.S., along a line stretching from Oregon to Texas, then down to the Yucatan Peninsula in Mexico, through most countries in Central America, to Colombia and finally Brazil. In the rest of North America and in parts of Central and South America, it will be seen as a partial solar eclipse. The eclipse begins at 16:10 UTC and ends at 19:49 UTC. Observers elsewhere in North America will see a partial eclipse of varied magnitude.

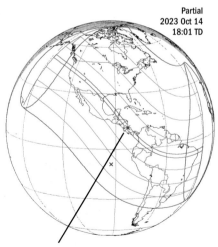

Partial
2023 Oct 14
18:01 TD

point of greatest eclipse

Thousand Year Canon of Solar Eclipses
© 2014 by Fred Espenak

The Northern Lights

The Sun appears to be a bright, unchanging orb in the sky. However, the Sun is anything but unchanging: As we know, it is a ball of constant activity, and that activity has a large influence here on Earth.

One such activity is a solar flare, which is often followed by a coronal mass ejection (CME) that sends particles speeding along the solar wind. Solar flares can reach Earth and disrupt radio transmissions. If Earth is in the path of a CME, the particles can travel down our magnetic field lines toward the poles; this creates the beautiful Northern and Southern Lights, or **aurora borealis** and **aurora australis**, respectively.

As mentioned earlier, the Sun goes through an average 11-year cycle of activity during which it experiences a solar minimum and a solar maximum. In the latter part of 2020, we began a new Solar Cycle, and as the cycle continues, we can expect to see more of the Northern Lights. It's worth noting that over the past few cycles the Sun has been less active than in the past.

Auroras come in different shapes and sizes. They can be steady, moving or rapidly pulsating. They also come in an array of colors, depending on how the particles interact with molecules at different altitudes. They are also hard to predict and can be difficult to see. Catching them is a special treat, even for experienced astronomers. Be aware that to the unaided eye the auroral colors are usually muted, as the light is not strong enough to stimulate our color vision. The bright colors seen in photographic reproductions of auroras show up when long exposures are used.

Green is the most common color and occurs when particles interact with oxygen molecules at an altitude of roughly 100 to 300 kilometers (62 to 186 miles). Between 300 to 400 kilometers (186 to 248 miles) the oxygen molecules produce a red color, and below 100 kilometers (62 miles) the particles interact with nitrogen — producing a pink color.

While the interaction of these solar particles can produce a beautiful light show, they can also be destructive.

One of the most powerful events from a CME was called the Carrington Event. In 1859, English astronomers Richard Carrington and Richard Hodgson were the first to ever witness a solar flare. But when the particles reached Earth, they disrupted telegraph systems in North America and Europe, in some reports even setting equipment on fire. The Northern Lights were visible as far south as Honolulu and the Southern Lights as far north as Santiago, Chile.

A massive CME, like the one that caused the Carrington Event, has a real chance of disrupting GPS and satellite communications and destroying electrical grids. As a result, space agencies have been launching more satellites and probes to the Sun to monitor space weather, and power companies have been developing contingency plans to ensure the next powerful solar eruption doesn't cause such damage.

The Northern Lights over Saskatchewan

The Planets

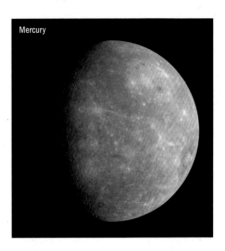

Mercury

As the planets (including Earth) revolve around the Sun, there are a number of notable events that take place, each of which has a specific name and meaning.

The inner planets Mercury and Venus never stray far from the Sun. When either planet is farthest from the Sun in the western evening sky, it is said to be at **greatest eastern elongation**; when either is farthest from the Sun in the eastern morning sky, it is said to be at **greatest western elongation**. These are the best opportunities to view the inner planets, though Venus can be seen at other times.

The term **conjunction** has a specific technical meaning — that is, when two objects have the same right ascension — but amateur astronomers also use the term casually when there is a close approach of two or more celestial objects.

Then there are **oppositions**: when the outer planets (Mars, Jupiter, Saturn, Uranus, Neptune) are opposite the Sun in the sky in right ascension. Around opposition dates, the outer planets appear closer, brighter and larger in diameter — great for telescopic viewing. The Sky Month-by-Month section (pages 42–113) lists eastern and western elongations, conjunctions and oppositions throughout 2023.

Mercury is a place of extremes. As the smallest of our Solar System's eight planets at one-third the size of Earth, Mercury is the closest planet to the Sun, sitting an average distance of 58 million kilometers (36 million miles) away. The planet, however, is only the second-hottest one in our Solar System. Mercury also orbits the Sun faster than any other planet and has the longest solar day, lasting 176 Earth days. Mercury has no moon.

The planet's surface warms to 427 degrees Celsius (801 degrees Fahrenheit) at the peak of its "noonday" heat; on the far side, its temperature drops down to a chilly –163 degrees Celsius (–261 degrees Fahrenheit). Mercury's coldest locations are deep in the shadows of the craters, where the Mercury Surface, Space Environment, Geochemistry and Ranging (MESSENGER) spacecraft mission to the small planet discovered ice. Temperatures in crater shadows are at –183 degrees Celsius (–297 degrees Fahrenheit).

Due to the planet's proximity to the Sun, Mercury is a challenging target to observe. When it is visible in Earth's sky, it appears in a narrow window of time in the dawn twilight just before sunrise or in the evening twilight just after sunset, depending on whether it is west or east of the Sun in the sky. To observe Mercury, one needs a low horizon unobstructed by trees or buildings, as it is typically only 15 to 25 degrees away from the Sun when visible.

Venus, the brightest planet in the sky, is a spectacular sight to see. Known as both the "morning star" and the "evening star," Venus is similar in size to Earth. But it's definitely not a place you'd like to visit. The planet is covered in dense clouds and, as a result, suffers from a runaway greenhouse effect. The planet's surface has temperatures of 460 degrees Celsius (860 degrees Fahrenheit) almost constantly.

A view of the Martian landscape from NASA's
Perseverance rover in March 2021

Venus spins in the opposite direction than the other planets, but very slowly. A day on Venus is longer than its year: one day takes 243 Earth days, while one year is 225 Earth days. Its orbit is, on average, 108 million kilometers (67 million miles) from the Sun. Being shrouded in white clouds, there is nothing much to see on Venus, but binoculars and small telescopes reveal that Venus passes through phases, similar to the Moon. (Mercury also shows phases, but they are more difficult to see, as the planet's disk appears so small.)

In 2023, Venus can be found low in the evening sky after sunset, where it is joined by Jupiter until the beginning of March. On March 1, the pair of planets will be less than a quarter of a degree apart. Venus will remain in the evening sky until roughly the end of July.

There is no other planet that fascinates humans as much as **Mars**. In 1894, American astronomer Percival Lowell thought he could see deep canals carved through its surface that he believed were evidence of intelligent life. Later observations with spacecraft saw no evidence of these canals.

Mars is the most explored planet in our Solar System, and some even like to joke that it's the only planet with an entirely robotic "civilization" of Earth-made machines.

The orbiters, landers and rovers dispatched to the Red Planet have taught us a lot about its ancient past. Though rocky, dusty and seemingly devoid of any life, some evidence suggests Mars once had an ocean of water covering most of its northern hemisphere. Scientists sometimes cite the existence of water to suggest the planet was once habitable, though we have yet to find proof that life actually ever thrived there.

Mars is about half the size of Earth, with a day very close to Earth's 24-hour day. It lies roughly 228 million kilometers (142 million miles) away from the Sun. Small telescopes (with a 70 mm diameter and below) will show the red disk of Mars; a larger, good-quality telescope at high magnification will show surface features such as ice caps, dark basins and light deserts.

In December 2022, Mars was the closest it has been since 2018, so the brightest the Red Planet will be in 2023 will be in January.

Jupiter is the king of our planetary system, at 11 times wider than Earth. This gas giant boasts the most spectacular storm formation in our Solar System, called the Great Red Spot. Jupiter orbits 772 million kilometers (479 million miles) from the Sun and a single day is about 10 Earth hours long. It orbits the Sun in about 12 Earth years.

Jupiter is the second-brightest planet in our night sky and is also one of the most enjoyable planets to observe. With a modest telescope, you can easily see the cloud bands in its atmosphere. But you can even enjoy the planet through a pair of binoculars; if you watch the planet every night, four of its 79 confirmed moons — Ganymede, Io, Europa and Callisto — can be seen changing positions night after night.

The Great Red Spot (GRS) is a massive, swirling egg-shaped storm that has been shrinking for reasons that are unclear to astronomers. In the 1800s, the GRS was estimated at 41,000 kilometers (25,476 miles) along its long axis. NASA's Juno spacecraft measured the GRS at 16,350 kilometers (10,159 miles) in width on April 3, 2017. Several astronomy apps predict the best time on any night for viewing the GRS through a telescope.

Jupiter

During the evenings from January to March, Jupiter can be found in Pisces, which will be fairly low in the sky. It then returns to the early morning sky in June, but low near the horizon. By mid-August, it once again returns to the night sky, rising in the east just after midnight.

With its magnificent rings, **Saturn** is often considered the jewel of our Solar System. With a pair of binoculars, Saturn's pancake shape is evident. Look through a telescope, even a modest one, and its fine rings are clearly visible.

Saturn is nine times wider than Earth and 1.4 billion kilometers (886 million miles) away from the Sun. This gaseous planet has the second-shortest day in the Solar System, lasting roughly only 10.7 hours, while its orbit takes 29.4 Earth years. The planet also has more than 80 moons, though its largest, Titan, is the only one easily viewed through a modest telescope.

Saturn is the second-largest planet in the Solar System. It's not the only planet to have rings (Jupiter, Uranus and Neptune all have ring systems), but it has by far the most spectacular ring system we can see. Saturn's rings are made up of billions of pieces of ice that range from tiny dust-sized grains to chunks as big as a house. The ring system is roughly 282,000 kilometers (175,000 miles) across but only about 9 meters (30 feet) tall. There are several rings, mostly close to each other, but the main rings are called A, B, and C. The gap between the outer ring (A) and the first inner ring (B) is called the Cassini Division, following the discovery by Italian astronomer Giovanni Cassini.

In 2023, Saturn can be found low in the west after sunset until the end of January. For a beautiful sight, on January 22, look to the horizon after sunset and you'll see a beautiful pairing of Venus and Saturn, which will be less than a quarter degree apart. By May, Saturn rises in the east in the early morning hours;

Saturn

and by the middle of August you can find the marvelous ringed planet rising in the east just after sunset.

Uranus is the faintest of the planets visible to the naked eye (in dark-sky conditions), and the third-largest planet overall. This giant ice planet has 13 faint rings and 27 small moons. But most interestingly, Uranus rotates on its side, at nearly a 90-degree angle.

In 1781, Uranus was the first planet to be discovered with the aid of a telescope by astronomer William Herschel. However, Herschel first believed it was either a star or a comet. It was confirmed as a planet two years later by Johann Elert Bode.

The planet is four times the size of Earth. A day on Uranus takes roughly 17 Earth hours, with one orbit taking 84 Earth years. Uranus orbits on average 2.9 billion kilometers (1.8 billion miles) from the Sun.

Like it did in 2022, Uranus once again finds itself in Aries, reaching opposition on November 13 at magnitude + 5.6.

At an average distance of 4.5 billion kilometers (2.8 billion miles) from the Sun, **Neptune** is the farthest planet in our Solar System, taking about 165 Earth years to complete one orbit. Like Uranus, Neptune is roughly four times larger than Earth, and it, too, has a faint ring system.

Neptune also holds the distinction as being the windiest planet in our Solar System, with clouds of frozen methane being whipped across the planet at 2,000 kilometers (1,200 miles) per hour. A day on Neptune is about 16 Earth hours long.

In January and February, Neptune hangs out in Aquarius then shifts to Pisces for the rest of 2023. It reaches opposition on September 19 at magnitude +7.8.

Deep Sky Objects

The Universe has many amazing sights to behold, including stunning clusters of stars, nebulae and swirling galaxies.

There are two main types of star clusters: **open** and **globular**. Globular clusters are old star systems at the edge of spiral galaxies that can contain anywhere from thousands to millions of stars, packed in a close, roughly spherical form and held together by gravity.

Two beautiful globular clusters you can see from the Northern Hemisphere include Messier 13, found in the constellation Hercules (a hero from Greek mythology), and Messier 3 in the constellation Canes Venatici (the hunting dogs).

Open clusters are found on the **galactic plane**, the plane on which most of a galaxy's mass lies. They contain anything from a dozen to hundreds of stars, but the stars are more spread out than globular clusters. Perhaps the most famous open star cluster, the Pleiades (Messier 45) can be spotted with the naked eye and through light pollution.

As well, there are **nebulae** — clouds of dust and gas. These are considered "stellar nurseries," as eventually the gas and dust will coalesce into new stars and potential stellar systems with planets, moons and, possibly, life. One of the most famous and easily visible is the Orion Nebula (Messier 42), found in the winter constellation Orion.

Some nebulae are leftover dust and gas from a **supernova**, an exploding star. There are also **emission nebulae**, which are clouds of interstellar gas excited by nearby stars that emit their own light at optical wavelengths. **Planetary nebulae** are cloudy remnants left over from stars that shed their gas and dust late in their lives. And finally, there are **dark nebulae**, interstellar clouds that are so dense they obscure the light of the objects behind them.

Messier 13, a globular cluster, seen here with Mars

Messier Catalog

The Messier Catalog is a register of 110 objects in the night sky, including open and globular clusters, nebulae, galaxies and one double star. The catalog was started by Charles Messier in the 18th century. Messier was chiefly interested in finding comets, and the catalog is his list of non-comet objects that he observed. You can find a list of all the objects in the Messier Catalog on pages 114–117.

The Pleiades (Messier 45)

The Orion Nebula (Messier 42)

Galaxies

Galaxies come in many different shapes and sizes. There are four main types, however: elliptical, spiral, barred spiral and irregular. The diagram below shows Edwin Hubble's scheme for classifying galaxies, colloquially known as the "tuning fork" because of its shape.

Elliptical galaxies seem somewhat disorganized and look roughly egg-shaped. Shapes range from almost circular (E0) to very elliptical (E7).

Spiral galaxies have both a large central bulge and a thin disk of stars. Much like elliptical galaxies, these also range from tight spirals (Sa) to more diffuse (Sd).

Barred spiral galaxies are similar to spirals, but they have visible "arms" or bars near the center, and they range from tight (SBa) to diffuse (SBd).

There are also disk galaxies that don't have spiral arms. These are called **lenticular (lens-shaped) galaxies**. They are classified as S0.

The most common galaxy is the spiral, accounting for more than 75 percent of galaxies in the visible Universe. Our galaxy, the Milky Way, is believed to be a barred spiral. Our closest neighboring spiral galaxy is the Andromeda Galaxy (Messier 31), which is easily visible through binoculars or with the naked eye in dark skies.

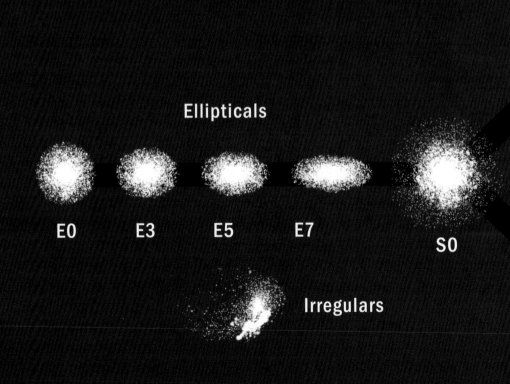

Ellipticals

E0   E3   E5   E7

S0

Irregulars

The Andromeda Galaxy (Messier 31)

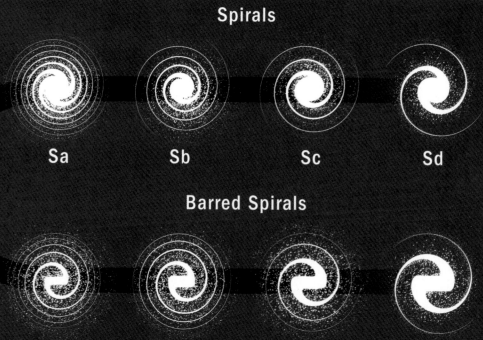

Spirals

Sa   Sb   Sc   Sd

Barred Spirals

SBa   SBb   SBc   SBd

The Sky Month-by-Month

Introduction

The following pages are your guide to the sky for every month in 2023. Each month features a calendar of events, information about Moon phases, sky charts facing south and north and descriptions of interesting objects to target.

Monthly Events

This section summarizes the events of each month, including Moon phases, conjunctions between the Moon and planets, conjunctions between planets and other planets, eastern and western elongations, meteor shower peaks and much more.

All times are listed in UTC, or Coordinated Universal Time, which is the standard time used by astronomers throughout the year. The map shown below will guide you on how to calculate your local time from the time shown in UTC. It's important to note that most of North America will observe Daylight Saving Time (DST) between March 12, 2023, and November 5, 2023. Between these two dates, you will need to add an additional hour to your local time.

You'll notice some of the times listed fall during daylight, when observation is likely impossible. However, you can use this date

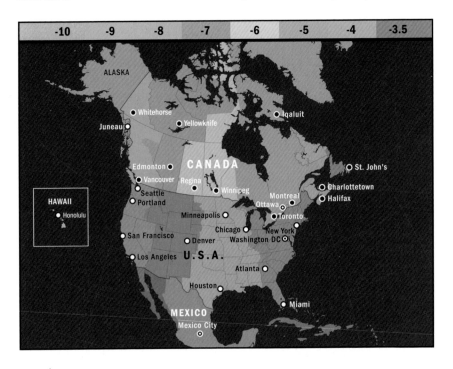

and time as a guide for observing an event on either the preceding night or the following night. Depending on your location, objects might also be below the horizon at the stated time, so again, use this calendar as a guide for which night these events can be observed. Similarly, the coordinates given as the distance between celestial objects may not reflect exactly what you will see in the night sky. This is particularly true for close approaches that take place during the day, local time. For example, the calendar of events might say Mercury is 1.5 degrees south of Jupiter at 21:00 UTC, but by the time Mercury and Jupiter become visible to an observer in Eastern Time living in Toronto, Canada, that separation might have increased to 2.4 degrees.

This section also features a calendar layout with the Moon phases for each day shown, as well as a description of interesting Moon events for the month. You might notice some discrepancies between the coordinates given in these descriptions and those listed in the table of events. Because the Moon moves in right ascension so quickly, conjunctions between the Moon and planets as listed in the table may appear different for North American observers, and the separation between the objects may be more than suggested. The coordinates in the descriptions have been altered to more closely reflect the view from North America, but of course there may be more variation based on when the objects are visible to you in the night sky.

The Moon phases shown in the calendar are based on the day they occur in UTC time. For some observers in North America, that might mean the ideal time to watch, for example, the full Moon rising would be the night before.

Sky Charts and Descriptions

Each month features two sky charts, one facing south and the other facing north. These charts highlight select constellations, stars, clusters, nebulae, planets and galaxies visible in the night sky. While these sky charts are an accurate representation of the sky, it should be noted that factors like light pollution, smoke, cloud cover and so on can affect how much you see, and some of the fainter stars shown on the charts may not be visible.

The charts are drawn at a latitude of 45 degrees north. Observers north or south of this latitude will see slightly more of the northern or southern sky, respectively. The charts show the sky at 10:00 p.m. local standard time on the 15th of each month. You will also have the same view of the sky at 11:00 p.m. local standard time at the beginning of each month and at 9:00 p.m. local standard time at the end of each month. We note these times on the sky charts and also include Daylight Saving Time in parentheses between March and November. The planets shown on the sky charts will move slightly relative to the stars over the course of the month.

The preceding month's charts can be used for sky viewing two hours earlier, and the following month's charts can be used for sky viewing two hours later. So, for example, if you wanted to view the sky at 8:00 p.m. local time in February, you would refer to January's sky chart. If you're referring to a previous or later month's sky chart, note that the positions of the planets will not be correct.

January

January's Events

It's the beginning of a new year and a great time for astronomy — if you can brave the colder weather. January's night sky has wonderful binocular targets for everyone to enjoy, including the Orion Nebula, the Pleiades and other star clusters.

The month starts off with a decent meteor shower: the Quadrantids. This shower, with a rate of about 120 meteors an hour, can produce some spectacular fireballs. It began on December 27 and peaks on the night of January 3 to 4.

In the morning sky, keep an eye out for Mercury in the east, low on the horizon. The first planet from the Sun reaches greatest elongation west on January 30, and your best chance of sighting it is before sunrise during the last 10 days of the month.

Calendar of Events

Day	Time (UTC)	Event
01	22:00	Uranus 0.7°S of Moon
03	01:24	Pleiades 2.5°N of Moon
03	19:35	Mars 0.6°N of Moon
04		Quadrantid meteor shower peak
04	19:59	Earth at perihelion: 0.9833 astronomical units (AU)
06	23:08	Full Moon
07	13:40	Pollux 2.1°N of Moon
08	09:19	Moon at apogee: 406,500 km (252,587 mi.)
15	02:10	Last quarter of the Moon
18	09:32	Antares 2.1°S of Moon
21	20:53	New Moon
21	20:58	Moon at perigee: 356,600 km (221,581 mi.)
22	21:53	Saturn 0.4°N of Venus
23	07:22	Saturn 3.8°N of Moon
23	08:20	Venus 3.4°N of Moon
26	02:00	Jupiter 1.8°N of Moon
28	15:19	First quarter of the Moon
29	04:00	Uranus 0.9°S of Moon
30	05:59	Mercury 25°W of Sun (greatest western elongation)
30	07:21	Pleiades 2.6°N of Moon
31	04:24	Mars 0.1°N of Moon

The Moon This Month

SUN	MON	TUES	WED	THURS	FRI	SAT
1	2	3	4	5	6 Full Moon	7
8	9	10	11	12	13	14
15 Last Quarter	16	17	18	19	20	21 New Moon
22	23	24	25	26	27	28 1st Quarter
29	30	31				

The new year begins with a Moon-Uranus conjunction. The Moon, which will be roughly 85 percent illuminated, will lie within 2 degrees of our seventh planet (depending on where you live), though you may need a large telescope to see it (you can try binoculars).

On January 3, the almost-full Moon joins two noticeably red objects in the sky. One is the bright red star Aldebaran in the constellation Taurus; the other is Mars. On that night, Mars and the Moon will be roughly 3 degrees apart.

The peak of the Quadrantids falls on the night of January 3 to 4, which isn't ideal, as the Moon will be nearly fully illuminated. Though this shower doesn't typically produce meteors with long trains, it is known for fireballs, so observers can still keep an eye out for these bright meteors.

When it comes to the planets, Jupiter remains high in the southwest for the month. On January 25, it will be joined by a slim crescent Moon. Mars can be found high in the east after sunset and remains in a good position in Taurus all night.

On January 30, the Moon and Mars join once again, around 2 to 3 degrees apart (depending on the time and location).

The full Moon falls on January 6, with the new Moon on the January 21.

The Moon hits apogee (the farthest from Earth in its monthly orbit) on January 8 and perigee (the closest to Earth in its monthly orbit) on January 21.

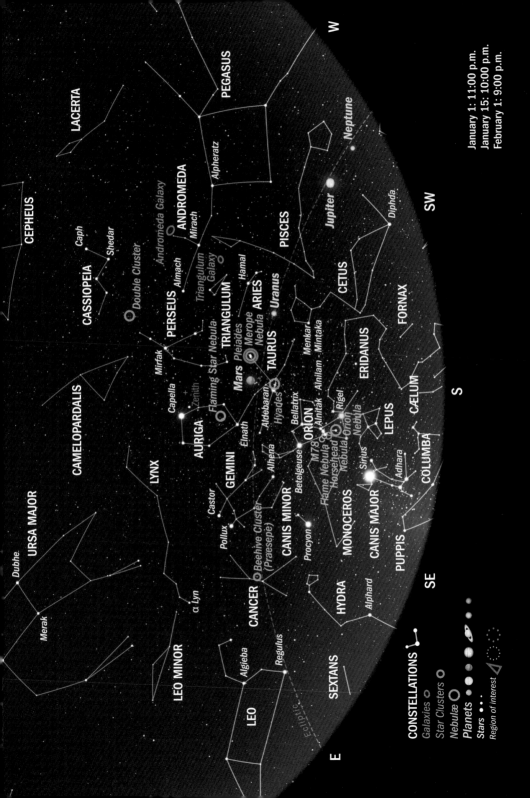

W

January 1: 11:00 p.m.
January 15: 10:00 p.m.
February 1: 9:00 p.m.

SW

S

SE

E

Neptune

Jupiter

Diphda

PEGASUS

Alpheratz

LACERTA

CEPHEUS

CASSIOPEIA

Caph

Shedar

Double Cluster

Andromeda Galaxy

ANDROMEDA

Mirach

PERSEUS

Almach

Triangulum Galaxy

TRIANGULUM

Hamal

ARIES

Uranus

PISCES

CETUS

Menkar

FORNAX

Mirfak

Flaming Star Nebula

Pleiades

Merope Nebula

TAURUS

Aldebaran

Hyades

Elnath

Zenith

Capella

AURIGA

Mars

CAMELOPARDALIS

LYNX

Castor

Pollux

GEMINI

Alhena

Bellatrix

Betelgeuse

ORION

Alnitak · Alnilam · Mintaka

Rigel

Orion Nebula

M78

Flame Nebula

Horsehead Nebula

ERIDANUS

CÆLUM

LEPUS

COLUMBA

Adhara

Sirius

CANIS MAJOR

PUPPIS

MONOCEROS

Procyon

CANIS MINOR

Beehive Cluster (Praesepe)

CANCER

α Lyn

URSA MAJOR

Dubhe.

Merak

LEO MINOR

Algieba

Regulus

LEO

SEXTANS

Ecliptic

HYDRA

Alphard

CONSTELLATIONS

Galaxies ⊘

Star Clusters ○

Nebulæ ○

Planets ● ● ●

Stars • • •

Region of interest △ ○⋮

Beehive Cluster (Messier 44)

Highlights in the Southern Sky

The highlight of the winter sky is always the most recognizable constellation, **Orion (the Hunter)**. You can make out its most prominent feature: the three belt stars (from left to right) **Alnitak, Alnilam** and **Mintaka**. The entire constellation is rich with nebulosity, clouds of dust and debris. Around Mintaka is the **Horsehead Nebula (Barnard 33)**, as well as the **Flame Nebula (NGC 2024)**.

Below the belt stars is one of the most famous nebulae, the **Orion Nebula (Messier 42)**. The region is visible with the unaided eye, even from light-polluted cities. Try viewing it through binoculars.

Another beautiful nebula is the lesser-known **Messier 78**. This is a reflection nebula — a nebula that is illuminated by a nearby star — and it can be found roughly 2 degrees above Alnitak and lies 1,600 light-years from Earth. Observers can best see this nebula in large binoculars — it will look like a fuzzy patch — or telescopes.

You may also notice a particularly bright star to the east of Orion's left foot. That would be the unmistakable **Sirius**, the brightest star in our sky. Sirius, or the "dog star," is part of **Canis Major (the Great Dog)** and has a dim white-dwarf companion known as the "Pup."

Almost at the zenith you can find Mars together with **Aldebaran**, the bright-red star in **Taurus (the Bull)**. If you want to enjoy a particularly beautiful sight, pull out the binoculars to take a look. There you will see the open star cluster the **Hyades (Melotte 25)**. The cluster is named after the five daughters of Atlas, a Titan from Greek mythology.

The **Pleiades (Messier 45)**, another open star cluster, can also be found nearby in Taurus, about 7 degrees from Mars. In Greek mythology, the Pleiades are half-sisters to the Hyades.

If you enjoy open star clusters, another favorite of astronomers is the **Beehive Cluster (Messier 44)**, one of the nearest to Earth. The cluster, also known as **Praesepe (the Manger)**, is 600 light-years away. It is home to about 1,000 stars that are just 600 million years old — young compared to our own "senior" Sun, which clocks in at 4.5 billion years old.

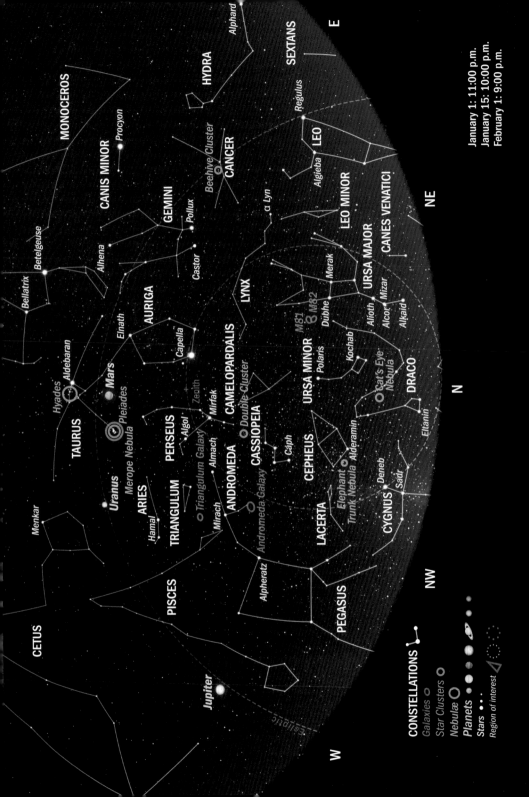

E

SEXTANS

Alphard

HYDRA

Regulus

LEO

CANCER

Beehive Cluster

CANIS MINOR

Procyon

MONOCEROS

LEO MINOR

Algieba

NE

URSA MAJOR

CANES VENATICI

Merak

Dubhe

Alioth
Mizar
Alcor
Alkaid

M81
M82

Castor

Pollux

GEMINI

Alhena

LYNX

α Lyn

AURIGA

Capella

Elnath

CAMELOPARDALIS

Double Cluster

URSA MINOR

Polaris

Kochab

N

Cat's Eye
Nebula

DRACO

Eltanin

Betelgeuse

Bellatrix

TAURUS

Hyades
Aldebaran

Mars

Pleiades
Merope Nebula

Uranus

ARIES

Hamal

PERSEUS

Mirfak

Algol

Almach

ANDROMEDA

CASSIOPEIA

Caph

Andromeda Galaxy

Mirach

TRIANGULUM

Triangulum Galaxy

CEPHEUS

Alderamin

Elephant
Trunk Nebula

CYGNUS

Deneb

Sadr

LACERTA

Zenith

CETUS

Menkar

PISCES

Alpheratz

PEGASUS

NW

W

Jupiter

ECLIPTIC

January 1: 11:00 p.m.
January 15: 10:00 p.m.
February 1: 9:00 p.m.

Messier 82

Highlights in the Northern Sky

Slinking across the northern sky, beginning just above the horizon stretches **Draco (the Dragon)**, one of the largest constellations in the sky. It often gets overlooked for the other more prominent constellations, but it contains some lovely objects, including the **Cat's Eye Nebula (NGC 6543)**, a planetary nebula that you will need a telescope to observe.

Planetary nebulae are produced from dying stars as they shed their gas and dust at the end of their lives and were first identified by astronomer William Herschel. The surrounding gas of the Cat's Eye Nebula is expanding at a speed of 16.4 kilometers per second (10.2 miles per second).

Of course, the most well-known object in the northern sky is the **Big Dipper**. It isn't a constellation; rather it is an asterism, or a group of stars within a constellation that forms a recognizable pattern. The Big Dipper lies within the constellation of **Ursa Major (the Great Bear)**.

Two wonderful galaxies also lie within Ursa Major: **Messier 81 (Bode's Galaxy)** and **Messier 82 (the Cigar Galaxy)**. M81 can be seen from dark-sky sites through binoculars or a small telescope, and M82 is in the same field of view. The two are often photographed together as they are extremely close in the night sky.

February

February's Events

You may have noticed that the nights are beginning to get a little shorter now as the daylight hours increase. But the cloud-covered nights are beginning to clear, which is good news for those who love to gaze up at a star-covered sky.

If you have a pair of binoculars, you can continue to enjoy several treats in Orion, which is prominent in the south. Just use your binoculars to explore — you are sure to find many intriguing targets that you can then try to identify using the provided star charts.

February's also a great time to gaze to the north, particularly if you have a small telescope, as two of the most well-known galaxies — Messier 81 and 82 — are in a great location high up in the sky.

The Andromeda Galaxy is also high in the west, as well as Perseus's Double Cluster.

And, when it comes to planets, Mars features prominently high in the southern sky while Venus and Jupiter dominate the western evening sky. Neptune and Venus have a very close approach (about half a degree) on the evening of February 15 that you will need a telescope to appreciate.

Calendar of Events

Day	Time (UTC)	Event
03	19:47	Pollux 2.1°N of Moon
04	08:55	Moon at apogee: 406,500 km (252,587 mi.)
05	18:29	Full Moon
13	16:01	Last quarter of the Moon
14	18:09	Antares 1.9°S of Moon
15	12:00	Neptune 0.01°N of Venus
19	09:06	Moon at perigee: 358,300 km (222,637 mi.)
20	07:06	New Moon
22	07:57	Venus 2.1°N of Moon
22	21:58	Jupiter 1.2°N of Moon
26	14:42	Pleiades 2.2°N of Moon
27	08:06	First quarter of the Moon
28	04:32	Mars 1.2°S of Moon

The Moon This Month

SUN	MON	TUES	WED	THURS	FRI	SAT
			1	2	3	4
5 Full Moon	6	7	8	9	10	11
12	13 Last Quarter	14	15	16	17	18
19	20 New Moon	21	22	23	24	25
26	27 1st Quarter	28				

Messier 81 and 82 are in a prime viewing location in February

The beginning of the month features a waxing gibbous Moon, illuminated at roughly 80 percent.

It will be at apogee on February 4 and at perigee on February 19. The full Moon falls on February 5, with the new Moon on February 20.

On February 22, a thin crescent Moon will lie just over 1 degree south of Jupiter, with Venus closer to the horizon just after sunset. The trio will be unmistakable, as both Jupiter and Venus are the brightest planets visible in our sky.

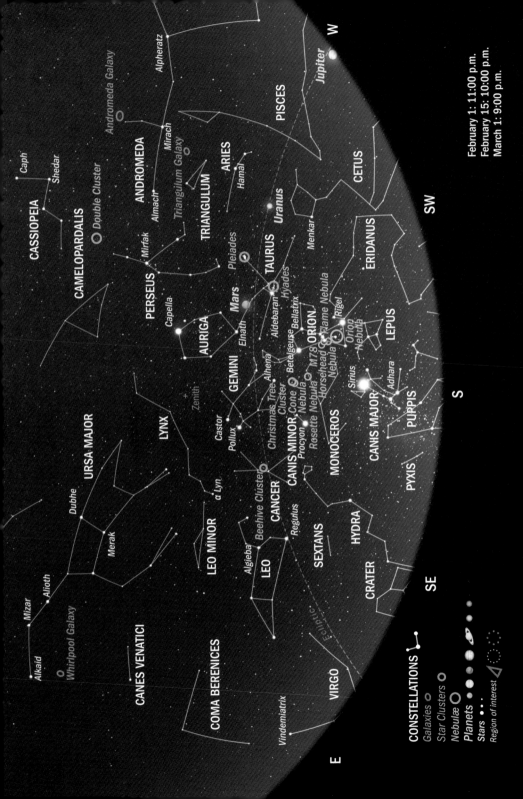

W

Jupiter

PISCES

Alpheratz

Andromeda Galaxy

Caph

Shedar

Mirach

CETUS

CASSIOPEIA

Double Cluster

Almach

ARIES

Triangulum Galaxy

Hamal

CAMELOPARDALIS

ANDROMEDA

February 1: 11:00 p.m.
February 15: 10:00 p.m.
March 1: 9:00 p.m.

Mirfak

TRIANGULUM

Uranus

PERSEUS

Pleiades

Menkar

SW

Capella

TAURUS

Elnath

Mars

Hyades

Aldebaran

ERIDANUS

AURIGA

Bellatrix

GEMINI

Alhena

Betelgeuse

ORION

Flame Nebula

Rigel

Zenith

Christmas Tree

Cone Nebula

Horsehead

Orion
Nebula

LEPUS

Pollux

Cluster

M78

Nebula

Castor

Procyon

Rosette Nebula

Adhara

S

LYNX

CANIS MINOR

Sirius

α Lyn.

MONOCEROS

CANIS MAJOR

URSA MAJOR

Beehive Cluster

PUPPIS

Dubhe

CANCER

Regulus

PYXIS

Merak

Algieba

SEXTANS

Alioth

LEO MINOR

LEO

HYDRA

SE

Mizar

LEO

Alkaid

Whirlpool Galaxy

CRATER

CANES VENATICI

COMA BERENICES

VIRGO

Vindemiatrix

E

CONSTELLATIONS

Galaxies ○

Star Clusters ○

Nebulæ ○

Planets

Stars • • •

Region of interest

Messier 78

Highlights in the Southern Sky

While **Orion (the Hunter)** is truly the richest target of the southern sky — with the **Orion Nebula (Messier 42)**, **Messier 78**, the **Flame Nebula (NGC 2024)** and **Horsehead Nebula (Barnard 33)** taking center stage — there is a plethora of other targets to enjoy.

To the left of Orion is **Monoceros (the Unicorn)**, which has two really interesting targets: the **Rosette Nebula (NGC 2237)** and the **Cone Nebula (NGC 2264)**.

The Rosette Nebula is a star-forming region that lies 5,200 light-years from Earth. It has a radius of about 65 light-years — much larger than the Orion Nebula, which is only 12 light-years in radius. But because it is much farther than the Orion Nebula (it sits just over 1,300 light-years away), it isn't as visible in the sky. However, at the Rosette Nebula's center is a rich open cluster of stars, NGC 2244, which is easily visible through binoculars.

The Cone Nebula is also home to another star-forming region, often referred to as the **Christmas Tree Cluster** (both are referred to as NGC 2264). The stars are easy to spot, but a large telescope is needed to see the Cone Nebula.

There are also two very prominent stars you may have noticed near Orion. Those would be **Castor** and **Pollux**, two of the brightest stars in **Gemini (the Twins)**, which represent the heads of the mythological Greek twins. Castor is just 51 light-years away and is part of a multiple-star system with three binary systems within it — meaning there is a total of six stars in all. Pollux is an orange-hued giant star that lies a bit closer to us, at 33 light-years away.

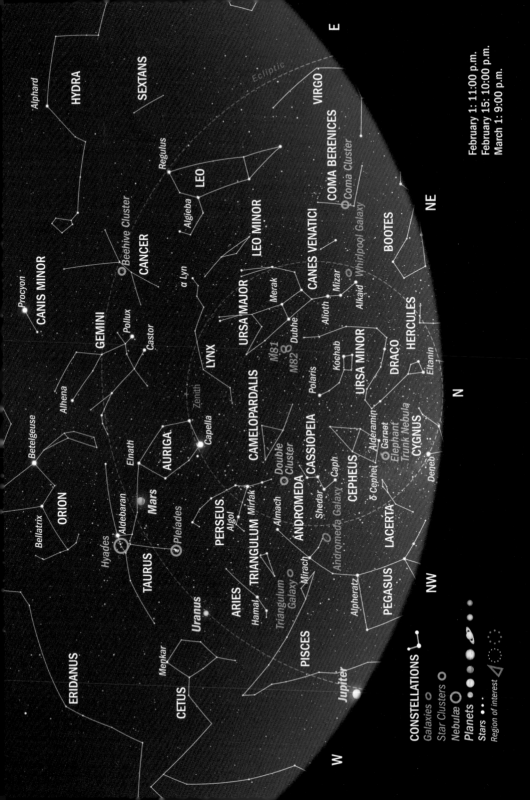

Messier 81

Highlights in the Northern Sky

If you're looking for due north, you can find it easily using the stars of the **Big Dipper**, which consists of seven bright stars in **Ursa Major (the Great Bear)**. Find the two stars at the top of the Big Dipper's "bowl": **Merak** to the right and **Dubhe** to the left. If you trace a line from Merak to Dubhe and beyond, you'll find **Polaris**, the brightest star in the **Little Dipper** asterism. In the Northern Hemisphere, Polaris is almost due north. If you were to set up a camera for a long exposure and point your lens at Polaris, you would notice the constellations moving around it as the night progresses.

Sticking close to the Big Dipper and Ursa Major, there are several galaxies in that area, including two of the most famous: **Messier 81 (Bode's Galaxy)** and **Messier 82 (the Cigar Galaxy)**. If you happen to get to a dark-sky site, try to find M81 through a pair of binoculars. Have patience, though — it can be tricky! While M82 is also in the same field of view, it is even more difficult to see.

There is also the **Whirlpool Galaxy (Messier 51)**, which lies near the first star in the Big Dipper's handle, **Alkaid**.

Perseus can be found high in the northwest with its pair of open clusters, appropriately dubbed the **Double Cluster (NGC 884 and NGC 869)**. From a dark-sky site, the pair can be seen with the naked eye. Grab a pair of binoculars and the display is even more spectacular.

March

March's Events

This month amateur astronomers around the world get excited for a very special event: the Messier Marathon. Observers can grab their binoculars or telescopes and scan the skies to find every one of the 110 Messier objects on a single night. For most of these objects, a pair of binoculars with a magnification of 70 millimeters or more can do the job. It's best to get to as dark a sky as you can.

Just after sunset on March 1, there is a conjunction of two of our brightest planets in the west, Venus and Jupiter. The pair will be less than a quarter of a degree apart. Mars is in a great place, remaining high in the sky this month. There are two good opportunities to spot the dim planet Uranus this month, when it is very close to brighter objects: On the evening of March 24, Uranus will be less than 2 degrees south of the crescent Moon; on the evening of March 31, Uranus and Venus pair up in a similar way. Try binoculars to find Uranus, but you may need a small telescope.

Finally, this month spring arrives on March 20, the vernal equinox.

Calendar of Events

Day	Time (UTC)	Event
02	04:15	Jupiter 0.5°N of Venus
03	02:10	Pollux 1.9°N of Moon
03	18:01	Moon at apogee: 405,900 km (252,215 mi.)
07	12:40	Full Moon
14	00:21	Antares 1.7°S of Moon
15	02:08	Last quarter of the Moon
19	15:16	Moon at perigee: 362,700 km (225,371 mi.)
19	15:20	Saturn 3.6°N of Moon
20	21:25	Vernal (spring) equinox
21	17:23	New Moon
22	20:00	Jupiter 0.5°N of Moon
24	10:28	Venus 0.1°N of Moon
25	01:00	Uranus 1.5°S of Moon
25	23:42	Pleiades 2°N of Moon
28	13:16	Mars 2.5°S of Moon
29	02:32	First quarter of the Moon
30	09:23	Pollux 1.7°N of Moon
31	06:00	Uranus 1.3°S of Venus
31	11:18	Moon at apogee: 404,900 km (251,593 mi.)

The Moon This Month

SUN	MON	TUES	WED	THURS	FRI	SAT
			1	2	3	4
5	6	7 Full Moon	8	9	10	11
12	13	14	15 Last Quarter	16	17	18
19	20	21 New Moon	22	23	24	25
26	27	28	29 1st Quarter	30	31	

On the night of March 2, you can find the Moon roughly 1.5 degrees away from Pollux, the brightest star in Gemini. And if you're up early on March 14, you can find the Moon near Antares, the bright red star of the summer constellation, Scorpius, which is now visible in the early morning hours in the south.

The Moon will be at apogee on March 3 and 31, and perigee occurs on March 19. (If it seems odd that the Moon could have two apogees or two perigees in a month, note that the Moon completes an orbit in about 27.6 days, compared to the longer calendar months of 28 to 31 days.)

The full Moon falls on March 7, with the new Moon on March 21 — just a day after the vernal equinox.

A waxing gibbous Moon

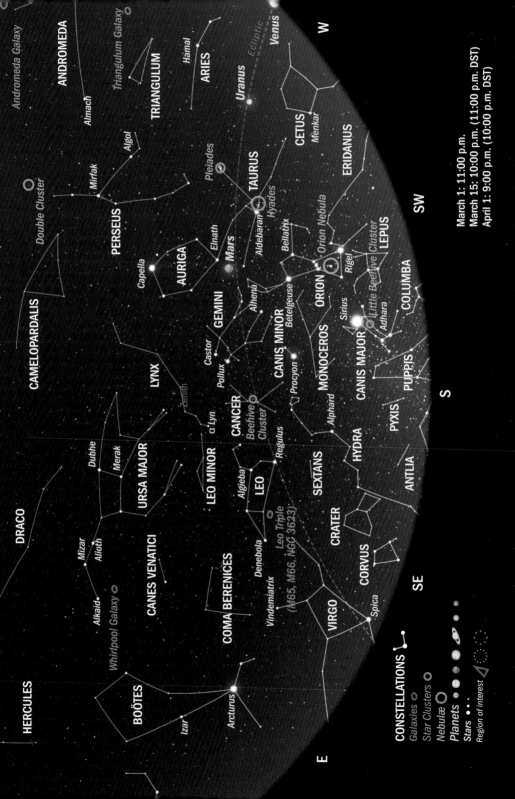

E

HERCULES

DRACO

Alkaid • Izar
Mizar
Alioth
Arcturus

BOÖTES

Whirlpool Galaxy ○

CANES VENATICI

COMA BERENICES

Vindemiatrix

Denebola

Leo Triple
(N65, M66, NGC 3623)

VIRGO

Spica

Algieba
LEO

Regulus

Dubhe
Merak

URSA MAJOR

LEO MINOR

α¹ Lyn.
Zenith

CANCER
Beehive Cluster

SEXTANS

CRATER

CORVUS

SE

CAMELOPARDALIS

LYNX

Castor
Pollux

GEMINI

Alhena

CANIS MINOR

Procyon

Alphard

HYDRA

ANTLIA

S

Double Cluster ○

PERSEUS

Mirfak

Algol

Capella

AURIGA

Elnath
Mars

Aldebaran

TAURUS

Bellatrix

Betelgeuse
ORION

Orion Nebula

Rigel

MONOCEROS

Sirius
CANIS MAJOR
Adhara

Little Beehive Cluster
LEPUS

PYXIS

PUPPIS

ANDROMEDA

Almach

Triangulum Galaxy

TRIANGULUM

Hamal

ARIES

Pleiades

Hyades

Uranus

Ecliptic

Venus

W

Andromeda Galaxy

CETUS

Menkar

ERIDANUS

COLUMBA

SW

March 1: 11:00 p.m.
March 15: 10:00 p.m. (11:00 p.m. DST)
April 1: 9:00 p.m. (10:00 p.m. DST)

CONSTELLATIONS ⊶
Galaxies ○
Star Clusters ○
Nebulæ ○
Planets ● ● ●
Stars • • •
Region of interest ◁

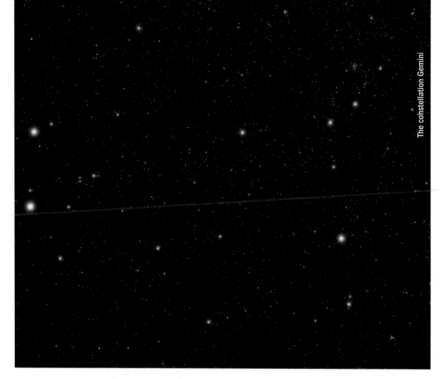

Highlights in the Southern Sky

Orion (the Hunter) is now beginning to dip to the south; his shield points to the western horizon as the nights go on.

Sirius is the most prominent object in the southern sky, as part of **Canis Major (the Great Dog)**. Though the constellation never gets particularly high up in the sky in the Northern Hemisphere, it's home to many star clusters and makes for a good binocular viewing. Look for the **Little Beehive Cluster (Messier 41)**, a bright, open star cluster found near Sirius.

And speaking of clusters, **Gemini (the Twins)** is high in the south – putting the **Beehive Cluster (Messier 44)** in a perfect location to see with binoculars, as it has risen higher in the sky over the past few months.

The longest of the 88 constellations, **Hydra (the Water Snake)** can be found stretching from the southeastern horizon up toward **Cancer (the Crab)**.

The Coma Cluster

Highlights in the Northern Sky

Cepheus, which looks like a little house, can be found in the northern sky. Nearby sits **Cassiopeia**, which looks like a sideways "W."

If you have a telescope, be sure to point it toward **Canes Venatici (the Hunting Dogs)**, **Coma Berenices (Berenice's Hair)** and **Virgo (the Maiden)**. The region is home to thousands of galaxies, called the **Coma Cluster**. The cluster lies about 330 million light-years from Earth and spans roughly 25 million light-years. It is also believed to be home to a supercluster of galaxies spanning some 200 million light-years.

If it's nebulae you like, plenty can be found in the northern sky this month, including the **Pacman Nebula (NGC 281)**, the **Wizard Nebula (NGC 7380)**, the **Cocoon Nebula (IC 5416)**, the **Heart Nebula (IC 1805)** and the **Soul Nebula (IC 1848)**. They are all visible through a small telescope and make wonderful photographic targets.

If you're going for the Messier Marathon, you'll find three objects near Cassiopeia: the **Andromeda Galaxy (Messier 31)**, **Messier 32** and **Messier 110** all clustered together. Andromeda, which is visible to the unaided eye away from light pollution, is fairly easy to spot. It lies between **Shedar** in Cassiopeia and **Mirach** in the constellation **Andromeda**.

April

April's Events

The warmer weather is beginning and the nights are getting shorter. However, this also means that the rich targets of the summer sky will rise earlier as the nights go on.

Mercury is at greatest elongation east (19.5 degrees) from the Sun on April 11 and should be visible low in the western evening twilight for several nights before and after. Venus is prominent in the western sky after sunset, though Jupiter is no longer visible. The red planet, Mars, is still high in the southern sky. Saturn is now beginning to rise in the east about an hour before sunrise. On April 15 and 16, the Moon will be very close to our Solar System's ringed planet.

There's also the Lyrid meteor shower this month, though it's not a strong shower. It only has a ZHR of 18, but on the bright side, it is known for producing some fireballs. The peak falls on the night of April 22 to 23.

As well, there is a rare hybrid solar eclipse (see page 29), but it will only be visible over the Indian Ocean, across Indonesia, Cambodia, Vietnam and Australia to the Pacific Ocean.

Calendar of Events

Day	Time (UTC)	Event
06	04:34	Full Moon
10	05:50	Antares 1.6°S of Moon
11	04:40	Pleiades 2.6°S of Venus
11	21:59	Mercury 19.5°E of Sun (greatest eastern elongation)
13	09:11	Last quarter of the Moon
16	02:22	Moon at perigee: 368,000 km (228,665 mi.)
16	03:47	Saturn 3.5°N of Moon
20	04:12	New Moon
20	04:17	Hybrid solar eclipse (not observable from North America)
22	09:14	Pleiades 2°N of Moon
23		Lyrid meteor shower peak
23	13:03	Venus 1.4°S of Moon
26	02:18	Mars 3.6°S of Moon
26	17:25	Pollux 1.7°N of Moon
27	21:20	First quarter of the Moon
28	06:43	Moon at apogee: 404,300 km (251,220 mi.)

The Moon This Month

SUN	MON	TUES	WED	THURS	FRI	SAT
						1
2	3	4	5	6 Full Moon	7	8
9	10	11	12	13 Last Quarter	14	15
16	17	18	19	20 New Moon	21	22
23	24	25	26	27 1st Quarter	28	29
30						

On April 22 and 23, the Moon will pass near Venus, and on April 25 the Moon will be just 2 degrees from Mars.

The Moon is at perigee on April 16 and at apogee on April 28. The full Moon falls on April 6, and the new Moon falls on April 20.

The Moon, Mars and Venus meet in the night sky

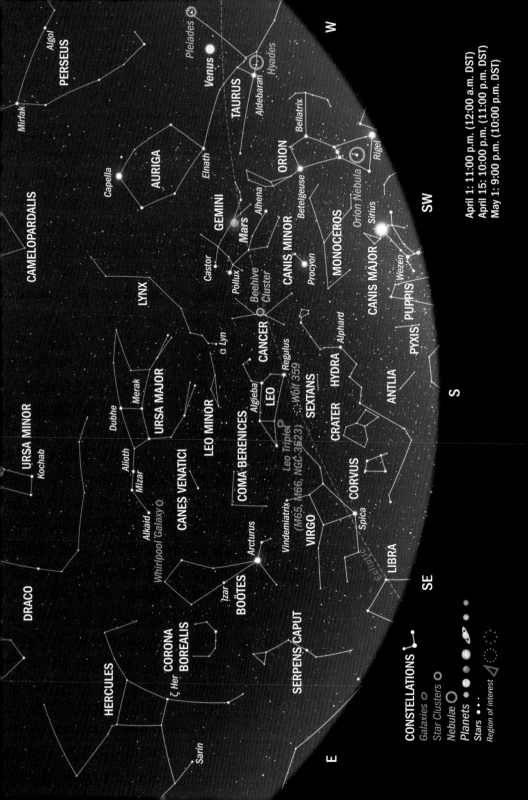

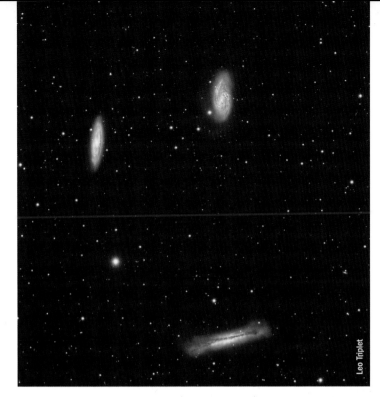

Leo Triplet

Highlights in the Southern Sky

Leo (the Lion) is now high in the south, lying beside **Cancer (the Crab)**. Two of the former constellation's brightest stars are **Regulus** and **Denebola**. You can also find a trio of galaxies close together, called the **Leo Triplet** near **Chertan**, a star that makes up the back end of Leo's body. It contains **Messier 65**, **Messier 66** and **NGC 3628**, the latter referred to as the **Hamburger Galaxy**.

Nearby you can also find the very dim star **Wolf 359**, which may sound familiar to fans of *Star Trek: The Next Generation* (it's the location where humanity fought the Borg). The small star is about the size of Jupiter and is the fifth-closest star to us. The red dwarf star is only visible through powerful telescopes.

Virgo (the Maiden), the second-largest constellation, is most known for being the location where the first exoplanets were discovered in 1992. Astronomers discovered the exoplanets around a pulsar, PSR B1257+12. Since then, more than 4,500 exoplanets have been discovered.

Meanwhile **Orion (the Hunter)**, the **Hyades (Melotte 25)** and the **Pleiades (Messier 45)** are making their way below the horizon in the west.

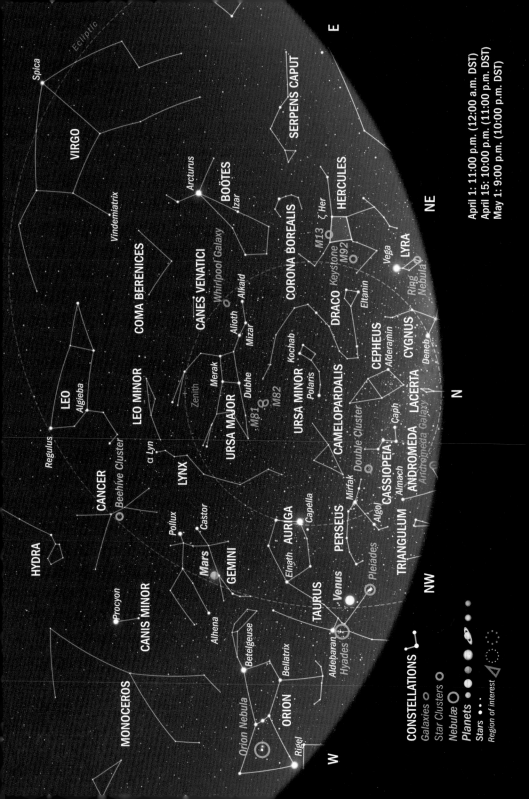

Messier 13

Highlights in the Northern Sky

Auriga (the Charioteer) can be found in the western sky, with its brightest star, **Capella**, shining brightly. The constellation contains three open clusters of stars easily found in binoculars: **Messier 36, Messier 37** and **Messier 38**. For good measure, slide over to the feet of **Gemini (the Twins)** to find another open cluster, **Messier 35**.

Boötes (the Herdsman) is now high in the east. The kite-like constellation — one of the largest of the 88 constellations — is most easily identifiable by finding its brightest star, **Arcturus**. The orange giant star is 25 times larger than our own Sun. While it may look like just a single star, it is actually a double, both of which are easily visible through a telescope.

You can now find **Hercules**, the fifth-largest constellation, in the northeastern sky, marked by the familiar "keystone." The most notable feature in the constellation is the bright globular cluster, **Messier 13**, which can be found in between the two stars known as **Zeta (ζ) Herculis** and **Eta (η) Herculis**. The cluster is about 25,000 light-years away from Earth and contains roughly 100,000 stars. Though best known for his discovery of a comet, Edmund Halley discovered this cluster in 1714. Nearby is another fainter globular cluster, **Messier 92**. Both are visible through binoculars.

May

May's Events

In May Mars remains the most prominent planet in the night sky, but you can also find Venus in the evening sky and Saturn in the morning. Jupiter is beginning to rise but remains close to the eastern horizon in the early morning hours.

This month there's a meteor shower, the Eta Aquariids, which began on April 19. Though it's best viewed from the Southern Hemisphere, on the peak night of May 6 to 7, you may be able to see 10 to 30 meteors an hour.

Calendar of Events

Day	Time (UTC)	Event
05	17:24	Penumbral lunar eclipse (not observable from North America)
05	17:34	Full Moon
07		Eta Aquariid meteor shower peak
07	12:35	Antares 1.6°S of Moon
08	11:23	Pollux 5°S of Mars
11	04:57	Moon at perigee: 369,300 km (229,472 mi.)
12	14:28	Last quarter of the Moon
13	13:04	Saturn 3.3°N of Moon
17	12:44	Jupiter 6.2°N of Mercury
17	13:15	Jupiter 0.8°S of Moon
18	01:34	Mercury 3.8°S of Moon
19	15:53	New Moon
23	12:08	Venus 2.4°S of Moon
24	01:36	Pollux 1.7°N of Moon
24	17:32	Mars 4.2°S of Moon
26	01:39	Moon at apogee: 404,500 km (251,345 mi.)
27	15:22	First quarter of the Moon
29	04:59	Mercury 24.9°W of Sun (greatest western elongation)
29	13:04	Pollux 4°S of Venus

The Moon This Month

SUN	MON	TUES	WED	THURS	FRI	SAT
	1	2	3	4	5 Full Moon	6
7	8	9	10	11	12 Last Quarter	13
14	15	16	17	18	19 New Moon	20
21	22	23	24	25	26	27 1st Quarter
28	29	30	31			

This month, the Moon dances with several planets. On May 13, there is a conjunction of the Moon and Saturn low on the horizon at the beginning of dawn. A few days later, on May 17, there is a conjunction of the Moon and Jupiter just before sunrise, but they will be very low on the horizon. Then, on May 22 and 23, you can enjoy a conjunction of a crescent Moon and Venus. The pair will be roughly 6 to 8 degrees apart, depending on where you live. On May 23, Mars will also be nearby. Finally, on May 24, the Moon pays Mars a visit in another conjunction. The two will appear 4 to 6 degrees apart.

The full Moon falls on May 5, with the new Moon on May 19. The Moon will be at perigee on May 11 and at apogee on May 26.

On May 23 Venus and Mars will lie near a crescent Moon

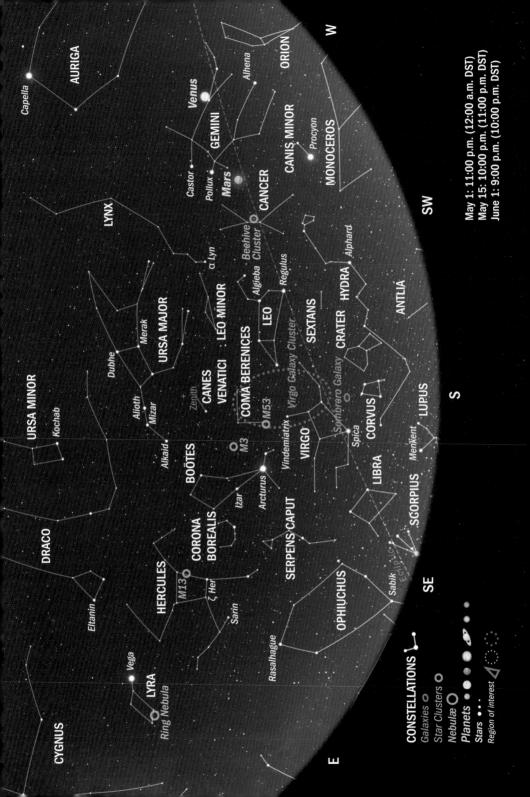

The Sombrero Galaxy (Messier 104)

Highlights in the Southern Sky

Directly to the south are two fairly faint constellations, **Corvus (the Crow)** and **Crater (the Cup)**. The pair lie low in the south, directly above **Hydra (the Water Snake)**, which is now fully visible (though also faint).

Leo (the Lion) and **Virgo (the Maiden)** are the most noticeable constellations in the south. Virgo's brightest star, **Spica**, shines in the sky; it is the 14th brightest star we can see in the night sky. Spica is a variable star, meaning its brightness changes over time. It is roughly 262 light-years away from Earth and is part of a binary system. Also in this constellation lies the **Virgo Galaxy Cluster**, home to roughly 2,000 galaxies, most of which can be seen in medium-sized telescopes.

In between Virgo and Corvus is the beautiful **Sombrero Galaxy (Messier 104)**. This is an edge-on galaxy with a prominent dust lane and is best seen through a telescope.

You can also find two beautiful globular clusters high in the east. The first, **Messier 3**, can be found in **Canes Venatici (the Hunting Dogs)**. This cluster is easily seen through large binoculars and will look like a fuzzy smudge. The cluster is believed to hold 500,000 stars and lies almost 34,000 light-years from Earth.

The second is **Messier 53**, found in **Coma Berenices (Berenice's Hair)**, also near the bright star **Arcturus** in **Boötes (the Herdsman)**. This globular cluster also holds at least 500,000 stars and is moving toward Earth at 112 kilometers per hour (70 miles per hour).

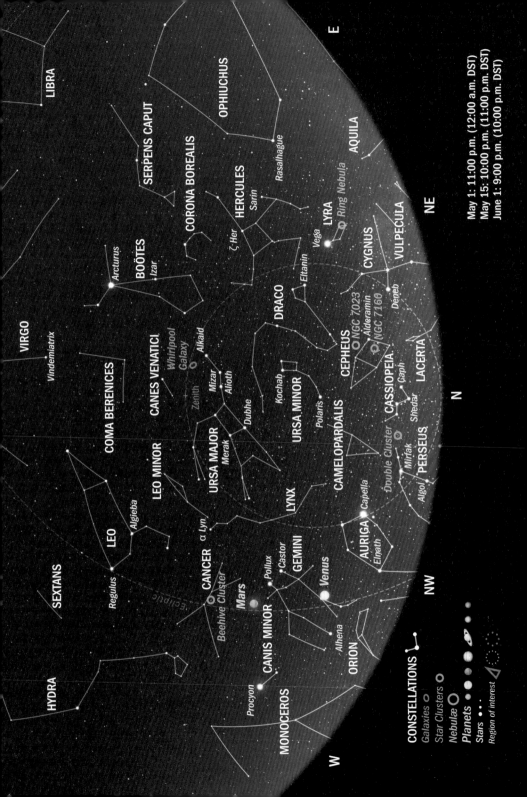

Iris Nebula (NGC 7023)

Highlights in the Northern Sky

Cassiopeia is now beginning to get quite low on the northern horizon. Nearby **Cepheus**, the little house, is beginning to tilt. You can use a pair of binoculars to find **NGC 7160**, a small open cluster found in the square of the house. Also found within Cepheus is the stunning **Iris Nebula (NGC 7023)**. This nebula is an emission nebula, which is a cloud of ionized gas. It is one of the rare nebulae that appears blue, making it a favorite subject for many astrophotographers.

Cygnus (the Swan) is now beginning to rise in the northeast. For a beautiful exploration of the night sky, grab a pair of binoculars and scan the region: it is a rich, starry landscape that is sure to amaze you.

Lyra (the Lyre), near Cygnus, is also placed nicely in the northwest. The constellation is easy to find due to its brightest star, **Vega**. The star is the fifth-brightest in the night sky and is relatively close by, just 25 light-years away. In about 14,000 years it will replace **Polaris** as our North Star. And if Vega sounds familiar, it was featured in *Contact*, a book by renowned astronomer Carl Sagan, which was turned into a movie in 1997.

June

June's Events

Warm weather is now upon us, which makes for more comfortable viewing, even though true darkness doesn't fall until much later in the night now.

The most prominent constellations, Sagittarius and Scorpius, are now beginning to rise in the southeast. You'll come to learn over the next few months just how many great binocular targets you can find in this region.

Jupiter still remains low in the east in the morning, and although Saturn is higher up it still remains visible only in the wee hours of the morning.

Mars and Venus are well-placed in the western sky after sunset. Venus is at greatest eastern elongation from the Sun (45.4 degrees) on June 4. Look for it low in the western sky after sunset. As the sky darkens, you might spot the planet Mars to the east and the twin stars Castor and Pollux to the west.

At the beginning of the month, Mars will be extremely close to the Beehive Cluster. On June 2, it will be at its closest and will make for a fantastic binocular target.

The summer solstice falls on June 21.

Calendar of Events

Day	Time (UTC)	Event
02	03:30	Beehive 0.1°S of Mars
03	21:19	Antares 1.6°S of Moon
04	03:42	Full Moon
04	11:01	Venus 45.4°E of Sun (greatest eastern elongation)
06	23:07	Moon at perigee: 364,900 km (226,738 mi.)
09	20:19	Saturn 3°N of Moon
10	19:31	Last quarter of the Moon
11	17:50	Pleiades 6.2°NW of Mercury
13	11:05	Beehive 0.5°N of Venus
14	06:33	Jupiter 1.6°S of Moon
16	00:47	Pleiades 2°N of Moon
18	04:37	New Moon
21	10:38	Beehive 4°S of Moon
21	14:58	Summer solstice
22	00:47	Venus 4.1°E of Moon
22	10:09	Mars 4.2°S of Moon
22	18:31	Moon at apogee: 405,400 km (251,904 mi.)
26	07:50	First quarter of the Moon

The Moon This Month

SUN	MON	TUES	WED	THURS	FRI	SAT
				1	2	3
4 Full Moon	5	6	7	8	9	10 Last Quarter
11	12	13	14	15	16	17
18 New Moon	19	20	21	22	23	24
25	26 1st Quarter	27	28	29	30	

On June 3, the Moon and Antares, the red supergiant star in Scorpius, will be roughly 2 degrees apart; this happens again on the night of June 30. The Moon will also swing by Saturn in the early morning of June 9 and 10. The waning crescent Moon joins Jupiter low in the east at dawn on June 14.

For a special treat, on June 21 look to the west, as the Moon will join Venus and Mars near the horizon in the evening twilight — a splendid sight in binoculars.

The Moon will be at perigee on June 6 and at apogee on June 22. The full Moon falls on June 4 and the new Moon falls on June 18.

The Summer Triangle asterism is prominent in the northern sky this month

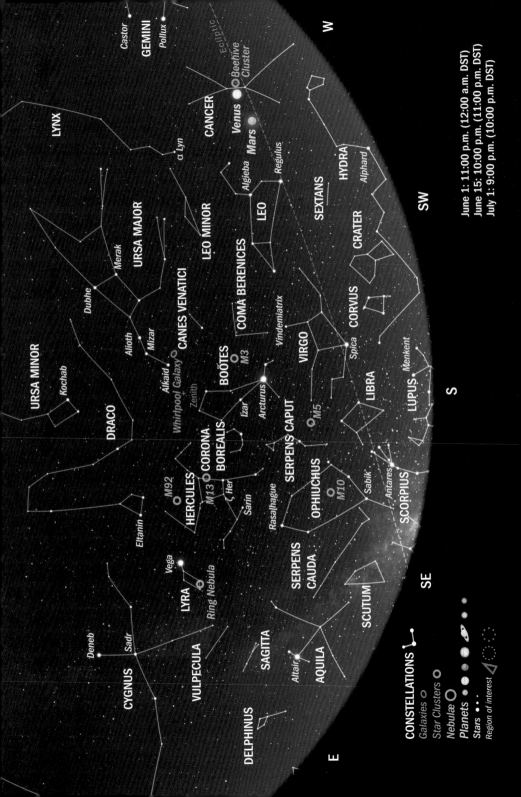

Highlights in the Southern Sky

The sky doesn't get truly dark until almost 10:00 p.m. or later in many parts of the Northern Hemisphere, which means there's not a lot of time for viewing unless you're willing to stay up quite late, but it's worth it. In northern parts of Canada, there is twilight from dusk until dawn!

In the southeast, you might notice a bright red star near the horizon. That's **Antares**, the brightest star of **Scorpius (the Scorpion)**. The name means "the rival of Mars," and the star is a red supergiant. Antares — which is 600 light-years from Earth — is 100 times the diameter of our Sun and 10,000 times more luminous. In Greek mythology, Scorpius stung Orion, a hunter who vowed to kill all the animals on Earth. Gaia,

the goddess of Earth rewarded Scorpius by placing him in the night sky where he can be found chasing Orion. That is why they are at opposite ends and are never seen together in the sky.

Ophiuchus (the Serpent Bearer) is also nearby in the southeast. At the center of this large constellation is **Messier 10**, another beautiful globular cluster that can be seen through binoculars.

Libra (the Scales) is also near Scorpius. Just above Libra, in the constellation **Serpens (the Serpent)**, is **Messier 5**, a beautiful globular cluster.

Boötes (the Herdsman) is now high in the south, with its brightest star, **Arcturus**, shining brightly.

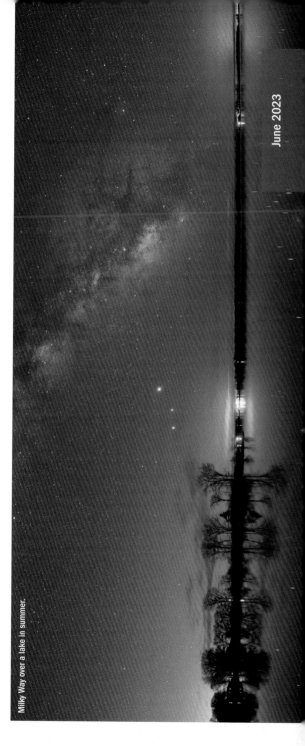

Milky Way over a lake in summer.

June 2023

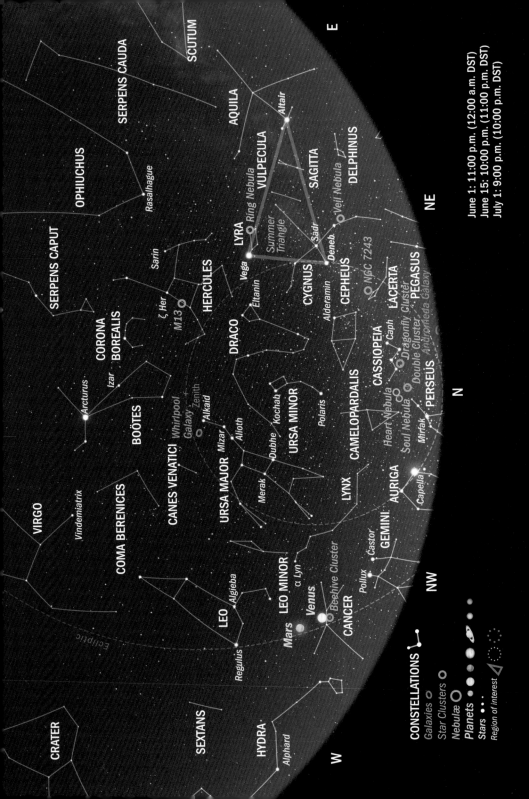

E

SCUTUM

SERPENS CAUDA

AQUILA

Altair

OPHIUCHUS

VULPECULA

DELPHINUS

Rasalhague

SAGITTA

Ring Nebula

Veil Nebula

NE

LYRA

Summer
Triangle

Sadr

SERPENS CAPUT

Sarin

Vega

Deneb

CEPHEUS

Eltanin

CYGNUS

NGC 7243

HERCULES

Alderamin

LACERTA

CORONA
BOREALIS

ζ Her

DRACO

PEGASUS

M13

Izar

Dragonfly Cluster

Andromeda Galaxy

Arcturus

Kochab

CASSIOPEIA

Caph

Double Cluster

PERSEUS

BOÖTES

Zenith

Whirlpool
Galaxy

Polaris

CAMELOPARDALIS

Heart Nebula

Soul Nebula

Mirfak

N

Alkaid

URSA MINOR

VIRGO

Alioth

Dubhe

Mizar

CANES VENATICI

URSA MAJOR

LYNX

AURIGA

Vindemiatrix

Merak

Capella

COMA BERENICES

Algieba

α Lyn

GEMINI

NW

LEO MINOR

Castor

LEO

Venus

Beehive Cluster

Pollux

Regulus

Mars

CANCER

SEXTANS

Ecliptic

HYDRA

CRATER

Alphard

W

June 1: 11:00 p.m. (12:00 a.m. DST)
June 15: 10:00 p.m. (11:00 p.m. DST)
July 1: 9:00 p.m. (10:00 p.m. DST)

CONSTELLATIONS

Galaxies ⌀

Star Clusters ⊘

Nebulæ ◯

Planets

Stars •••

Region of interest △

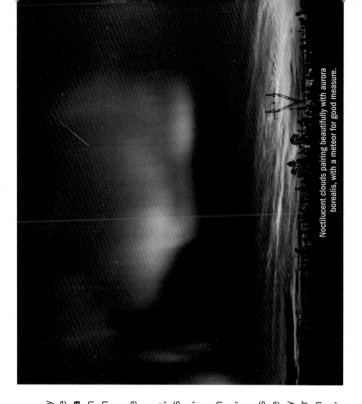
Noctilucent clouds pairing beautifully with aurora borealis, with a meteor for good measure.

Highlights in the Northern Sky

Cassiopeia is now seen as the "W" in the northern sky, and the nearby **Double Cluster (NGC 869 and NGC 884)** is due north but low on the horizon, together with the **Heart Nebula (IC 1805)** and the **Soul Nebula (IC 1848)**. A pair of binoculars will reveal many open star clusters in Cassiopeia, including the **Dragonfly Cluster (NGC 457)**. The constellation is a wonderful region to explore with binoculars.

The faint constellation **Camelopardalis (the Giraffe)** can also be found near Cassiopeia.

Cygnus (the Swan), is now clearly visible in the northwestern sky, as is **Lyra (the Lyre)** and **Aquila (the Eagle)**. The brightest stars in this trio of constellations make up the **Summer Triangle: Deneb** in Cygnus, **Vega** in Lyra and **Altair** in Aquila.

You can also find **Lacerta (the Lizard)**, a small, kite-like constellation near the "tail" of Cygnus. It has a beautiful open cluster, **NGC 7243**, which you can see in binoculars and small telescopes.

June is also a great month to try to observe **noctilucent clouds**, clouds that form high in the atmosphere and occur in the Northern Hemisphere beginning around the middle of May. While these clouds aren't entirely understood, it's believed that they consist of ice crystals high up in our atmosphere. Decades ago, these iridescent clouds used to only be seen in polar regions, but more recently they have been spotted farther south, even as far as Los Angeles.

July

July's Events

One of the best treats of the summer sky is the Milky Way, which stretches across the sky and is best seen in Scorpius and Sagittarius. There are so many things to observe with not only small telescopes, but also binoculars: nebulae, star clusters and more.

The best way to enjoy the summer sky is to get to a dark-sky location. One fun thing to do if you don't have binoculars or a telescope is to lie on a blanket and try to spot faint dots of light drifting across the sky. These would be satellites, and there are literally thousands up there. Many are visible to the unaided eye.

July is also a good month to look for star parties in your area. These events usually span a few nights, during which you can join like-minded people interested in astronomy. Don't have a telescope? No problem. Many people at these events are more than happy to share views of the night sky through their telescopes.

Venus and Mars are close together, sitting low in the western sky after sunset. Although they appear very close for several evenings, they never have a true conjunction; that is, they never lie along a north-south line in the sky relative to one another. Astronomers call this an appulse or quasi-conjunction. Venus has a true conjunction with Mercury on July 26, the two planets appearing about 5 degrees apart for several nights. On July 28, Mercury has an incredibly close conjunction with the star Regulus, the pair appearing only one-tenth of a degree apart. Use binoculars to pick them out in the twilight. Saturn rises in the east after midnight, which is good for those who are willing to stay up later to enjoy the shorter nights. Jupiter also rises close to 2:00 a.m.

Calendar of Events

Day	Time (UTC)	Event
01	06:48	Venus 3.6° from Mars (quasi-conjunction, see text)
03	11:39	Full Moon
04	22:28	Moon at perigee: 360,200 km (223,818 mi.)
06	08:59	Earth at aphelion: 1.0167 AU
07	03:05	Saturn 2.7°N of Moon
10	01:48	Last quarter of the Moon
11	21:18	Jupiter 2.3°S of Moon
17	18:32	New Moon
19	09:00	Mercury 3.5°S of Moon
20	06:56	Moon at apogee: 406,300 km (252,463 mi.)
21	04:00	Mars 3.6°S of Moon
25	22:07	First quarter of the Moon
27	11:00	Venus 5.1°N of Mercury

The Moon This Month

SUN	MON	TUES	WED	THURS	FRI	SAT
						1
2	3 Full Moon	4	5	6	7	8
9	10 Last Quarter	11	12	13	14	15
16	17 New Moon	18	19	20	21	22
23	24	25 1st Quarter	26	27	28	29
30	31					

On July 7, the Moon will be within roughly 3 degrees of Saturn very late at night. The waning crescent Moon passes Jupiter in the pre-dawn eastern sky on July 11 and 12. On July 19 and 20, low in the western evening sky just after sunset, the Moon begins another planetary dance: On July 19, the waxing crescent Moon forms a wide triangle with Venus and Mercury. Mars and the star Regulus are to the left, and the Moon joins them in a triangle on July 20. On the night of July 28, the waxing gibbous Moon appears 4 to 6 degrees to the left of Antares.

The Moon is at perigee on July 4 and at apogee on July 20, which is the same day the first humans set foot on the Moon back in 1969.

The full Moon occurs on July 3, while the new Moon falls on July 17.

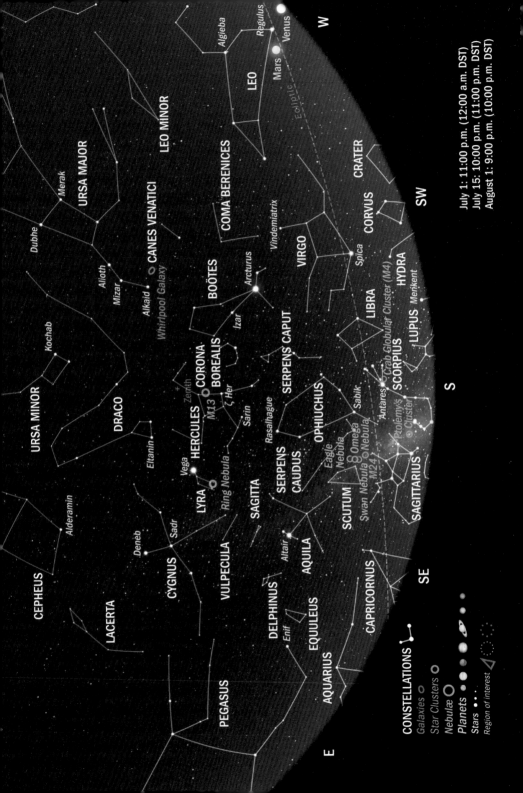

W

Venus
Regulus
Algieba
Mars

LEO

LEO MINOR

CANES VENATICI

URSA MAJOR
Merak
Dubhe
Alioth
Mizar
Alkaid
Whirlpool Galaxy

COMA BERENICES

BOÖTES
Arcturus
Izar

CRATER

SW

CORVUS

HYDRA
Menkent

VIRGO
Vindemiatrix
Spica

LIBRA

LUPUS

Crab Globular Cluster (M4)

SCORPIUS

S

Antares
Sabik
Ptolemy's
Cluster

OPHIUCHUS

Rasalhague

SERPENS CAPUT

CORONA
BOREALIS
ζ Her
Sarin
M13

Zenith

HERCULES

URSA MINOR
Kochab

DRACO
Eltanin

LYRA
Vega
Ring Nebula

SAGITTA

SERPENS
CAUDUS

SCUTUM
Eagle
Nebula
ω Omega
Swan Nebula Nebula
M24

SAGITTARIUS

SE

CEPHEUS
Alderamin

LACERTA

CYGNUS
Deneb
Sadr

VULPECULA

DELPHINUS
Enif

EQUULEUS

AQUILA
Altair

CAPRICORNUS

AQUARIUS

PEGASUS

E

CONSTELLATIONS
Galaxies ◦
Star Clusters ○
Nebulæ ◯
Planets • • ●
Stars • • ·
Region of interest ◁

July 1: 11:00 p.m. (12:00 a.m. DST)
July 15: 10:00 p.m. (11:00 p.m. DST)
August 1: 9:00 p.m. (10:00 p.m. DST)

Ecliptic

Highlights in the Southern Sky

July is a wonderful month to turn your eyes (or binoculars or telescopes) to the south, where the rich and diverse region around **Sagittarius (the Archer)** and **Scorpius (the Scorpion)** is in a great position in the sky. Grab your binoculars or telescope and get ready for a plethora of wonderful objects that are easily visible.

First, let's look at Sagittarius, which has the most interesting objects for amateur astronomers. A pair of binoculars will give you the best view of this constellation, which has an asterism that looks like a miniature teapot. To give an idea of just how impressive Sagittarius is, it holds the densest concentration of stars that can be seen through binoculars. There are an estimated 1,000 individual stars in a field of view.

The **Sagittarius Star Cloud (Messier 24)** resembles a cloud in the sky through binoculars. The cloud is 10,000 light-years from Earth and about 600 light-years wide.

Another target is the **Swan Nebula (Messier 17)**, also referred to as the **Omega Nebula**. This nebula is roughly 5,000 light-years from Earth and is one of the brightest star-forming regions of the Milky Way. It is easily visible through binoculars, but even a small telescope will reveal the nebulosity.

One of the best-known objects in this area is the **Eagle Nebula (Messier 16)**, which gained worldwide popularity when the Hubble Space Telescope released an iconic image of it entitled *The Pillars of Creation.*

Moving to Scorpius, **Antares** is the most prominent star in this summer constellation. Just beside it lies **Messier 4**. Both of these make wonderful binocular targets.

There's also **Ptolemy's Cluster (Messier 7)**, an open star cluster near the tail-end of Scorpius, almost straddling Sagittarius. The cluster lies just 980 light-years from Earth and is fairly easy to spot in dark skies.

Swan Nebula (Messier 17) and Eagle Nebula (Messier 16)

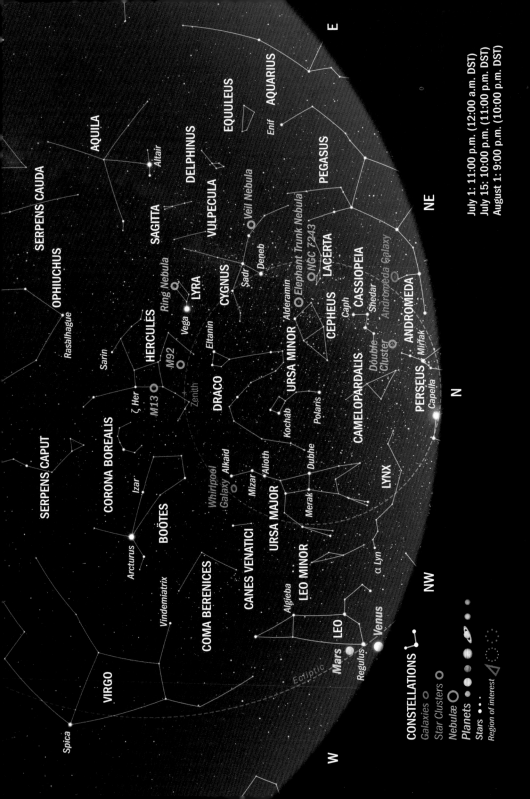

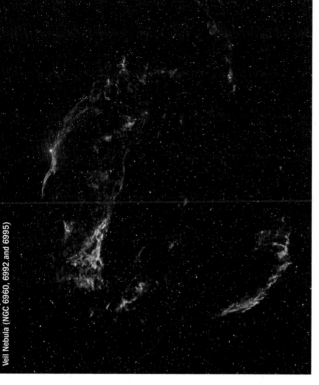

Veil Nebula (NGC 6960, 6992 and 6995)

Highlights in the Northern Sky

Cygnus (the Swan) is now well placed high in the northeastern sky — the swan stretching its wings as it flies into the night. One particularly beautiful sight in this constellation, which makes a great photographic target, is the **Veil Nebula (NGC 6960, 6992 and 6995)**. Though referred to as a single entity, this ancient supernova is made up of two parts: the Western Veil Nebula and the Eastern Veil Nebula. With its green-blue and red gases, the Veil Nebula is a favorite of many astrophotographers.

Hercules is now high in the north, which makes July a perfect time to seek out two great globular clusters, **Messier 13** and **Messier 92**.

The small constellation **Lacerta (the Lizard)** can be found near **Cepheus**. Though somewhat faint, it is also home to **NGC 7243**, a nice open cluster visible through binoculars. And speaking of Cepheus, you can also find here a lovely nebula that is another favorite of astrophotographers, the **Elephant's Trunk Nebula**. This nebula is a concentration of gas and dust within the much larger star-forming region known as **IC 1396**.

August

August's Events

This month Jupiter rises just before 5:00 a.m., which means late-night astronomers can take in the beautiful planet. If you gander at our Solar System's largest planet through binoculars you may notice Jupiter's four moons: Ganymede, Callisto, Io and Europa. If you can, take a look at Jupiter over a three-day period. You will notice the moons moving around Jupiter.

August is also time for one of the most anticipated meteor showers of the year, the Perseids. This shower can produce more than 100 meteors an hour at its peak night, which falls on the night of August 12 to 13. This year, the Moon is a waning crescent in the pre-dawn sky, so the dark skies will show off many meteors. The shower has been somewhat irregular over the past few years, with 2020 producing fewer meteors and 2021 peaking a night later than expected. If you can, grab a blanket and look for meteors the night before the peak, the night of and the night after.

For a few days, around August 13, Mars and Mercury appear about 5 degrees apart low in the west after sunset. They never undergo a true conjunction; that is, they never lie along a north-south line in the sky relative to one another. Astronomers call this an appulse or quasi-conjunction.

Calendar of Events

Day	Time (UTC)	Event
01	18:31	Full Moon
02	05:52	Moon at perigee: 357,300 km (222,016 mi.)
03	10:21	Saturn 2.4°N of Moon
08	09:41	Jupiter 3°S of Moon
08	10:28	Last quarter of the Moon
10	01:59	Mercury 27.4°E of Sun (greatest eastern elongation)
13	07:02	Mars 4.7° from Mercury (quasi-conjunction, see text)
13		Perseid meteor shower peak
13	21:36	Pollux 1.9° N of Moon
16	09:38	New Moon
16	11:55	Moon at apogee: 406,600 km (252,649 mi.)
18	23:06	Mars 2.4°SE of Moon
24	09:57	First quarter of the Moon
25	01:30	Antares 1.1°E of Moon
30	15:51	Moon at perigee: 357,200 km (221,954 mi.)
30	18:03	Saturn 2.4°N of Moon
31	01:35	Full Moon

The Moon This Month

SUN	MON	TUES	WED	THURS	FRI	SAT
		1 Full Moon	2	3	4	5
6	7	8 Last Quarter	9	10	11	12
13	14	15	16 New Moon	17	18	19
20	21	22	23	24 1st Quarter	25	26
27	28	29	30	31 Full Moon		

We start the month off with a full Moon and end it with a full Moon. The new Moon falls on August 16.

On August 8, you can find our lunar companion within roughly 3 degrees of Jupiter, rising in the east near midnight. The pair will remain in the sky all night.

On August 18, a thin crescent Moon will join Mars in the west after sunset.

On the evening of August 25, Antares appears within a degree of the Moon, and some lucky observers in central North America will get to watch the Moon pass in front of the bright star (contact your local astronomy club).

And this month perigees fall on August 2 and August 30, with apogee falling on August 16. (If it seems odd that the Moon could have two apogees or two perigees in a month, note that the Moon completes an orbit in about 27.6 days, compared to the longer calendar months of 28 to 31 days.)

A stunning halo surrounding the Moon

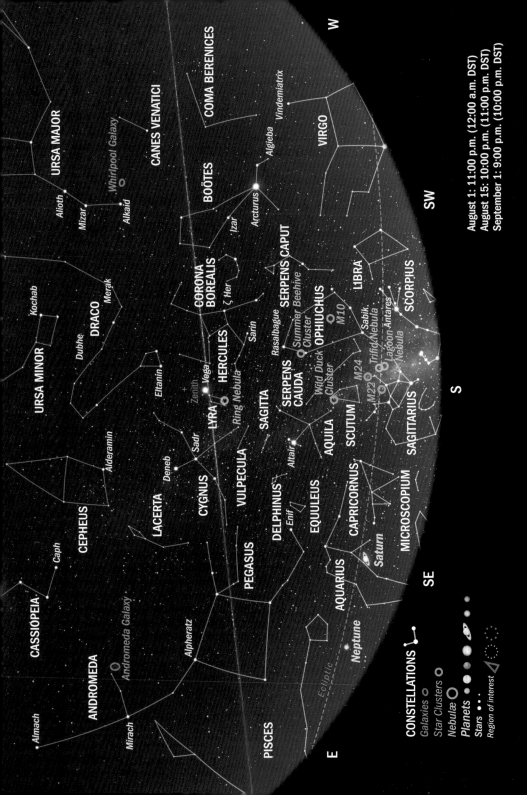

Lagoon Nebula (Messier 8) and Trifid Nebula (Messier 20)

Highlights in the Southern Sky

The rich region of the Milky Way still features prominently this month in the southern sky. However, **Scorpius (the Scorpion)** is beginning to hug the southern horizon, but **Sagittarius (the Archer)** is still in a fairly good place.

Ophiuchus (the Serpent Bearer) is high in the south, which makes August a great time to see **Messier 10**, a fairly bright globular cluster at Ophiuchus's heart. You can also use binoculars to find the open **Summer Beehive Cluster (IC 4665)** in the constellation.

Looking at Sagittarius, you can find the **Sagittarius Cluster (Messier 22)**, another beautiful globular cluster visible through binoculars.

There are also two beautiful nebulae in Sagittarius, the **Trifid Nebula (Messier 20)** and the **Lagoon Nebula (Messier 8)**. The Trifid Nebula is a large star-forming nebula that spans 42 light-years and sits 5,200 light-years from Earth. It is a favorite target of many astrophotographers and is easily visible through a small telescope. The Lagoon Nebula is another emission nebula and the largest and brightest within Sagittarius. It is believed to lie roughly 5,000 light-years from Earth. Through binoculars, it looks like a fuzzy patch, though detailed photographs reveal a bright-red region.

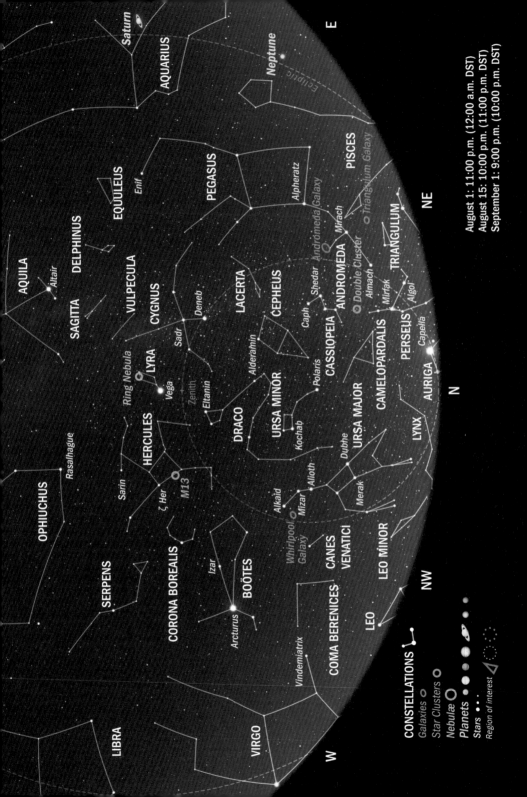

Whirlpool Galaxy (Messier 51)

Highlights in the Northern Sky

A favorite at star parties, the **Double Cluster (NGC 869 and NGC 884)**, between **Perseus** and **Cassiopeia**, rises higher in the northeast as the month goes on.

Ursa Major (the Great Bear) is now lower in the northern sky, but still contains wonderful objects, including the **Whirlpool Galaxy (Messier 51)**, found just off the tip of the last star of the handle, **Alkaid**.

Boötes (the Herdsman) has now swung to the west, though it is low in the west.

Cassiopeia is now well-placed in the northwestern sky. This means that August is a great month to see the closest spiral galaxy to our own Milky Way, the **Andromeda Galaxy (Messier 31)**. It can be easily found by finding the halfway point between **Shedar** in Cassiopeia and **Mirach** in **Andromeda** and then looking just north of that point. The galaxy can be seen by the naked eye in dark-sky locations or by using binoculars, where it appears as a faint fuzz. The Andromeda Galaxy is moving toward our own galaxy, and the pair will one day collide. Not to worry, though; this event won't happen for another 4.5 billion years.

September

September's Events

The colder weather is beginning to creep in, and the nights are getting longer.

The summer targets have now dipped low in the west and some even below the horizon, though Sagittarius remains visible low in the southwest until the end of the month.

Mars has now dipped below the horizon, but Saturn is visible all night. Venus can be found in the morning, and by the end of the month Mercury will be visible low in the eastern sky, undergoing its greatest western elongation (17.9 degrees) on September 22.

The Andromeda Galaxy (Messier 31) is a great target this month

Calendar of Events

Day	Time (UTC)	Event
04	19:44	Jupiter 3.4°S Moon
05	09:00	Uranus 2.8°S of Moon
05	19:25	Pleiades 1.3°N of Moon
06	22:21	Last quarter of the Moon
10	03:32	Pollux 1.7°N of Moon
12	15:42	Moon at apogee: 406,300 km (252,463 mi.)
15	01:40	New Moon
21	07:50	Antares 0.9°S of Moon
22	12:59	Mercury 17.9°W of Sun (greatest western elongation)
22	19:32	First quarter of the Moon
23	06:50	Autumnal (fall) equinox
27	01:25	Saturn 2.6°N of Moon
28	01:05	Moon at perigee: 359,900 km (223,631 mi.)
29	09:57	Full Moon

The Moon This Month

SUN	MON	TUES	WED	THURS	FRI	SAT
					1	2
3	4	5	6 Last Quarter	7	8	9
10	11	12	13	14	15 New Moon	16
17	18	19	20	21	22 1st Quarter	23
24	25	26	27	28	29 Full Moon	30

The Moon meets up with Jupiter late in the evening of September 4, and on September 5 it lies between the Pleiades and Jupiter, within a binocular field of Uranus. Before dawn on September 11, the waning crescent Moon appears with Venus low in the dawn sky, although widely separated by about 11 degrees.

On September 20 and 21, the Moon will lie not far from Antares in Scorpius, and on September 22, it will lie within Sagittarius.

On the evening of September 26, you can find Saturn and the Moon only about 2.5 degrees apart.

The new Moon falls on September 15, and the full Moon falls on September 29. It will be at apogee on September 12 and at perigee on September 28.

The Moon and Venus
over a peaceful lake

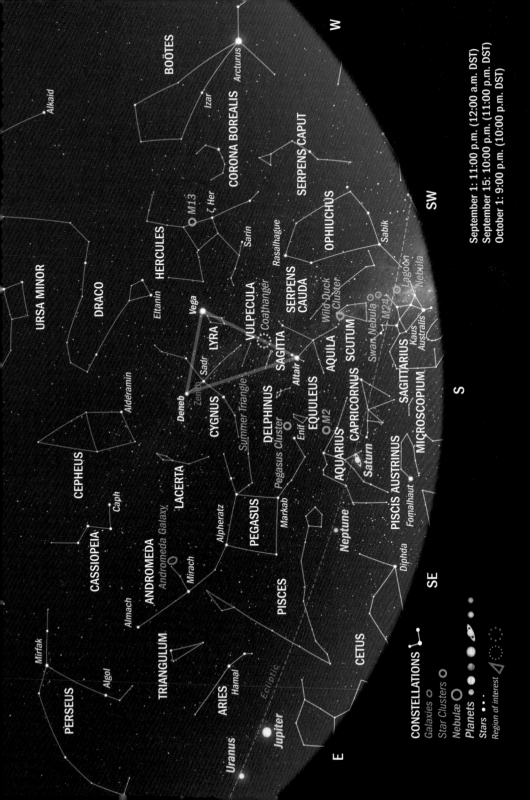

W

BOÖTES

Alkaid

CORONA BOREALIS *Arcturus*

Izar

SERPENS CAPUT

M13

ζ Her

HERCULES

Sarin

OPHIUCHUS

Rasalhague

SW

Sabik

URSA MINOR

DRACO

Eltanin

VULPECULA

Coathanger

SERPENS CAUDA

Vega

LYRA

Sadr

SAGITTA

Wild Duck Cluster

Swan Nebula

M24

Lagoon Nebula

CEPHEUS

Alderamin

Zeta

CYGNUS

Summer Triangle

Altair

AQUILA

SCUTUM

SAGITTARIUS

Kaus Australis

Deneb

DELPHINUS

Enif

EQUULEUS

Caph

Eltanin

LACERTA

Pegasus Cluster

M2

CAPRICORNUS

MICROSCOPIUM

S

CASSIOPEIA

Andromeda Galaxy

ANDROMEDA

Mirach

Alpheratz

PEGASUS

Markab

AQUARIUS

Saturn

PISCIS AUSTRINUS

Almach

TRIANGULUM

PISCES

Neptune

Fomalhaut

PERSEUS

Mirfak

Algol

ARIES

Hamal

Diphda

SE

CETUS

Uranus

Ecliptic

Jupiter

E

CONSTELLATIONS

Galaxies ○
Star Clusters ○
Nebulae ○
Planets
Stars • • •
Region of interest

September 1: 11:00 p.m. (12:00 a.m. DST)
September 15: 10:00 p.m. (11:00 p.m. DST)
October 1: 9:00 p.m. (10:00 p.m. DST)

Highlights in the Southern Sky

The **Summer Triangle** is still well-placed in the sky, with the bright stars **Deneb, Vega** and **Altair** forming the popular summer asterism.

Sagittarius (the Archer) is now low in the southwest, but a good pair of binoculars will still reveal many treats like the **Lagoon Nebula (NGC 6523)**, the **Sagittarius Star Cloud (Messier 24)** and the **Swan Nebula (Messier 17)**.

The **Wild Duck Cluster (Messier 11)** can be found in the small constellation **Scutum (the Shield)**. This compact open star cluster contains about 3,000 stars and lies about 6,200 light-years away from Earth. It is moving away from us at 22 kilometers per second (13.7 miles per second).

Pegasus is now prominent in the sky and is most easily recognizable by its large square. Using a pair of binoculars, you can also find two globular clusters nearby. **Messier 15**, also known as the **Pegasus Cluster,** can be found just off **Enif**, a star in Pegasus. Another is **Messier 2**, which lies in the nearby constellation **Aquarius (the Water Bearer)**.

To find a fun constellation, look to **Sagitta (the Arrow)**, within the Summer Triangle, just north of Altair. Just west of Sagitta is a wonderful asterism called the **Coathanger**, which is best seen through binoculars or a small telescope. The asterism looks like an upside-down coat hanger and is difficult to miss.

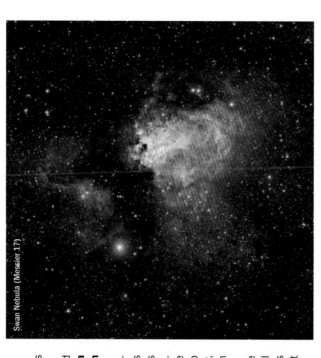
Swan Nebula (Messier 17)

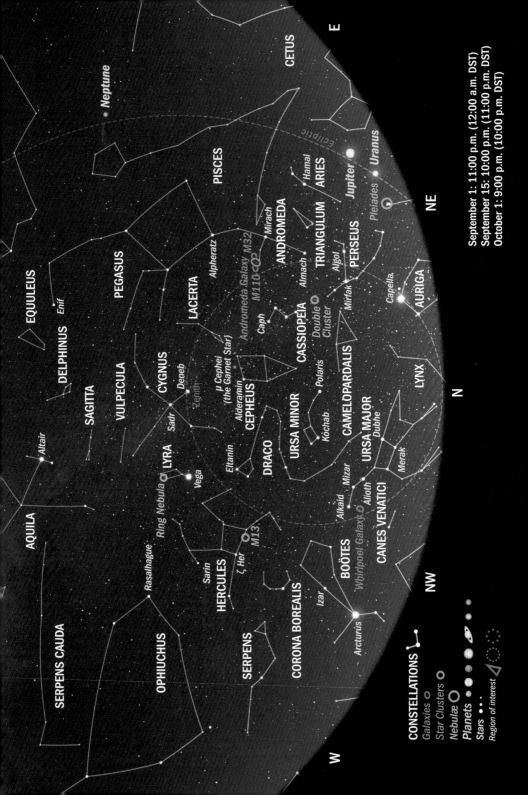

Highlights in the Northern Sky

Ursa Major (the Great Bear) is now low in the north. The bright star **Capella**, found in **Auriga (the Charioteer)**, is impossible to miss in the northeast.

Perseus is also in a good position in the northeast. The **Double Cluster (NGC 869 and NGC 884)** sits between **Perseus** and **Cassiopeia** and will be easily identifiable, especially under dark skies.

Another great dark-sky target is the **Andromeda Galaxy (Messier 31)**. While we often refer to the galaxy as a single object, there are two other galaxies in the same place, **Messier 32** and **Messier 110**.

Cepheus now rests high in the north. You can take the opportunity to find **Mu (μ) Cephei**, or "the Garnet Star," which gets its name from its deep reddish color.

Boötes (the Herdsman) can now be found in the west with its unmistakable bright star **Arcturus** shining brightly; however, over the course of the night, the constellation begins to dip below the horizon. Joining Boötes is **Hercules**, where you can view the impressive globular cluster **Messier 13**.

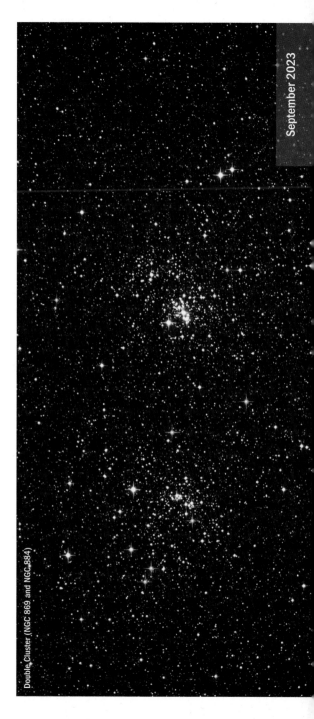

Double Cluster (NGC 869 and NGC 884)

October

October's Events

Longer, cooler nights are now upon us.

When it comes to the planets, both Saturn and Jupiter are well placed this month for viewing, as they are both prominent in the night sky. Venus is unmistakable in the morning sky, undergoing a greatest western elongation (46.4 degrees) on October 23.

And while most summer constellations are now gone, the Summer Triangle remains visible. It's a great place to explore with a pair of binoculars.

There's also an annular solar eclipse on October 14, which will be visible across the western United States. In the rest of North America, much of the eclipse will be seen as a partial solar eclipse.

Lastly, we have a meteor shower this month, the Orionids. It's not a big one, like the Perseid shower, but it can produce roughly 10 to 20 meteors an hour on its peak night, which falls on the night of October 21 to 22.

Calendar of Events

Day	Time (UTC)	Event
02	03:16	Jupiter 3.5°S of Moon
03	04:25	Pleiades 1.1°N of Moon
06	13:48	Last quarter of the Moon
07	10:23	Pollux 1.6°N Moon
09	06:10	Regulus 2.3°S of Venus
10	03:41	Moon at apogee: 405,400 km (251,904 mi.)
14	17:55	New Moon
14	18:00	Annular solar eclipse
18	13:17	Antares 0.9°S of Moon
22		Orionid meteor shower peak
22	03:29	First quarter of the Moon
23	23:14	Venus 46.4°W of Sun (greatest western elongation)
24	07:52	Saturn 2.8°N of Moon
26	02:53	Moon at perigee: 364,900 km (226,738 mi.)
28	20:15	Partial lunar eclipse
28	20:24	Full Moon
29	08:10	Jupiter 3.2°S of Moon
30	14:30	Pleiades 1.1°N of Moon

The Moon This Month

SUN	MON	TUES	WED	THURS	FRI	SAT
1	2	3	4	5	6 Last Quarter	7
8	9	10	11	12	13	14 New Moon
15	16	17	18	19	20	21
22 1st Quarter	23	24	25	26	27	28 Full Moon
29	30	31				

The month starts off with a close encounter between the Moon and Jupiter on October 1. The pair will be roughly 2 degrees apart. On October 2, the Moon lies just more than 1 degree from the Pleiades. In the dawn of October 9, in the east, look for the waning crescent Moon north of Venus, with the star Regulus to Venus's east.

On the nights of October 23 and 24, the Moon will swing by Saturn; the pair will be separated by 6 and 8 degrees respectively.

And on October 28, there's a partial lunar eclipse. However, there are a couple caveats: It will only be visible across the eastern Americas, Europe, Africa, Asia and Australia; also, only a small fraction of the Moon will be in the darkest part of Earth's shadow, the umbra. Most of the eclipse will be penumbral, which means the Moon is in the outer, fainter shadow. During penumbral eclipses, it's difficult to see any marked change in the brightness of the Moon.

On the night of October 28 to 29, the Moon passes around 4 degrees north of Jupiter.

The Moon is at apogee on October 10 and at perigee on October 26.

The new Moon falls on October 14, and the full Moon falls on October 28.

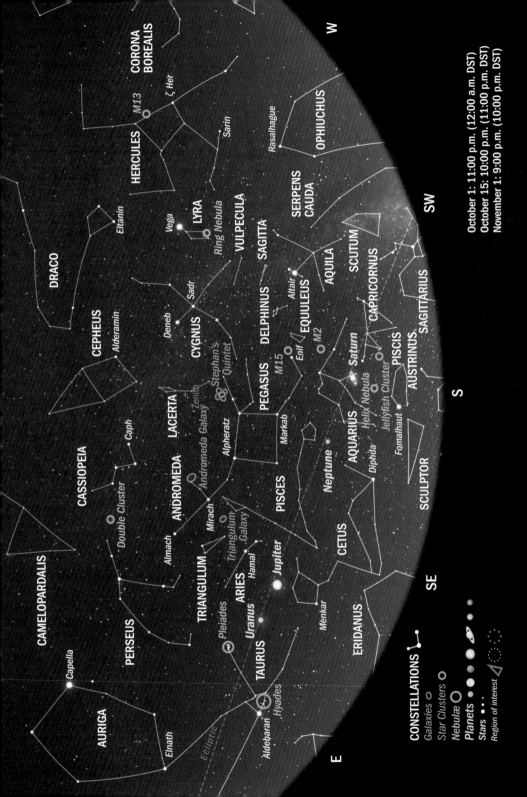

Stephan's Quintet

Highlights in the Southern Sky

Saturn is now in **Aquarius (the Water Bearer)** in the southern sky. An interesting target is just south of the planet: the **Helix Nebula (NGC 7293)**. The nebula is another bright, large planetary nebula and is just 650 light-years from Earth. It is one of the closest and brightest planetary nebulae. It can be found through binoculars under dark-sky conditions. Due to its appearance, it has also garnered the name "the Eye of God."

If you'd like a challenge, try grabbing a pair of binoculars and finding the **Jellyfish Cluster (Messier 30)** in **Capricornus (the Sea Goat)**. It's challenging to find, not only because it isn't a bright cluster, but also because the constellation — which isn't very bright in the first place — doesn't rise very high in the sky.

Pegasus is prominent in the sky with its giant square that's hard to miss. The constellation is also home to several interacting galaxies referred to as **Stephan's Quintet (NGC 7317, NGC 7318A, NGC 7318B, NGC 7319 and NGC 7320).** The first four of these galaxies lie roughly 300 million-light years from Earth and will one day merge into one giant galaxy. The last one — NGC 7320 — is much closer to Earth, at 39 million light-years away.

Low on the southern horizon is **Piscis Austrinus (the Southern Fish)**. It's not a bright constellation, but it does have a bright star that is very interesting, **Fomalhaut.** At one point, astronomers believed there was an exoplanet orbiting the star, however, by 2020 it had disappeared. Astronomers then proposed that the original object captured by the Hubble Space Telescope wasn't an exoplanet at all, but a cloud of dust left over from a collision of two icy bodies.

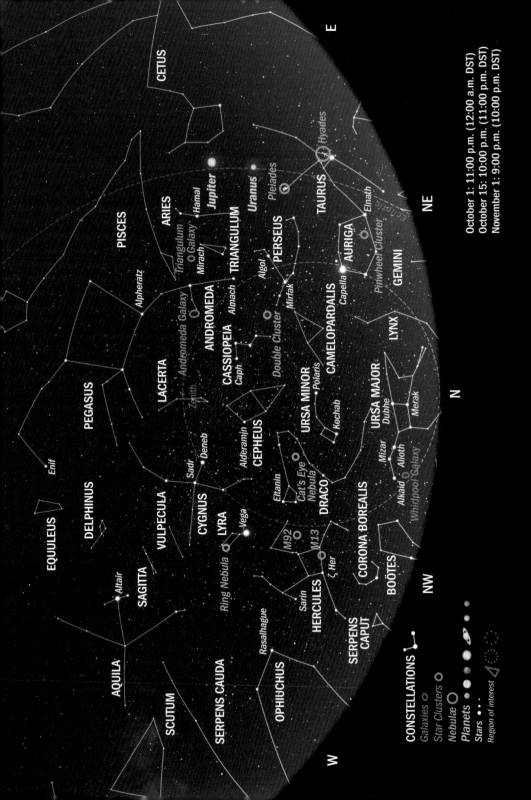

The Triangulum Galaxy (Messier 33)

Highlights in the Northern Sky

Draco (the Dragon) is now slinking across the northern sky, his arching back and tail stretching between **Ursa Minor (the Little Bear)** and **Ursa Major (the Great Bear)**. You can try finding the **Cat's Eye Nebula (NGC 6543)** in Draco, as it is placed very high in the sky this time of year.

To the west, there's **Lyra**, with its stunning bright star **Vega**. If you have a telescope, you can check out the **Ring Nebula (Messier 57)** within the constellation. The Ring Nebula is another planetary nebula left over from a dying star that has shed its outer shell, leaving only a small white dwarf behind. It lies roughly 2,300 light-years from Earth and is best seen through a large telescope.

Nearby, **Hercules** remains well-placed with its bright globular cluster **Messier 13** and the fainter **Messier 92**.

And looking to the northeast, the constellations **Perseus** and **Cassiopeia** are high in the sky, which means you still have a great view of the **Andromeda Galaxy (Messier 31)**.

Nearby there's a tiny constellation called **Triangulum**, where you can find the **Triangulum Galaxy (Messier 33)**. This galaxy is roughly half the size of our own Milky Way and is the third-largest member of our Local Group of galaxies — right behind the Andromeda Galaxy (the biggest) and the Milky Way (the second biggest).

November

November's Events

With summer far behind us, we now get ready to enjoy the treats of the winter night sky.

Orion — and all the beautiful targets this constellation has to offer — is now beginning to make an appearance in the southeast.

Jupiter and Saturn are still visible, though Saturn dips below the horizon later in the night. Venus remains a "morning star."

This month may be a good time to bundle up and try to catch some meteors. The Southern Taurids peak on the night of November 5 to 6 and the Northern Taurids peak on the night of November 12 to 13. As an added bonus, there's also the Leonid meteor shower, which peaks on the night of November 17 to 18. With all these showers peaking so close together, there's a good chance you'll see a "shooting star" this month.

Calendar of Events

Day	Time (UTC)	Event
03	18:31	Pollux 1.6°N of Moon
05	08:37	Last quarter of the Moon
06		Southern Taurid meteor shower peak
06	21:49	Moon at apogee: 404,600 km (251,407 mi.)
09	09:28	Venus 1.1°S of Moon
11	05:09	Spica 2.6°S of Moon
13		Northern Taurid meteor shower peak
13	09:27	New Moon
16	21:17	Antares 2.5°S of Mercury
18		Leonid meteor shower peak
20	10:50	First quarter of the Moon
20	14:02	Saturn 2.8°N of Moon
21	21:03	Moon at perigee: 369,800 km (229,783 mi.)
25	11:10	Jupiter 2.8°S of Moon
26	09:00	Uranus 2.8°S of Moon
27	00:02	Pleiades 1.2°N of Moon
27	09:16	Full Moon
29	10:29	Spica 4.2°N of Venus

The Moon This Month

SUN	MON	TUES	WED	THURS	FRI	SAT
			1	2	3	4
5 Last Quarter	6	7	8	9	10	11
12	13 New Moon	14	15	16	17	18
19	20 1st Quarter	21	22	23	24	25
26	27 Full Moon	28	29	30		

On November 9, a thin crescent Moon will be less than half a degree apart from Venus in the morning sky. Be sure to bring out the binoculars for this one. On November 20, the Moon will be roughly 4 degrees from Saturn in the south after sunset. In the late evening of November 25, Uranus and the waxing gibbous Moon are within a binocular field of each other.

The Moon is at apogee on November 6 and at perigee on November 21.

The new Moon is on November 13, and the full Moon falls on November 27.

The 2024 edition of *Night Sky Almanac* is in stores this month. Be sure to grab your copy, wherever books are sold.

The crescent Moon and Venus pair in the sky this month

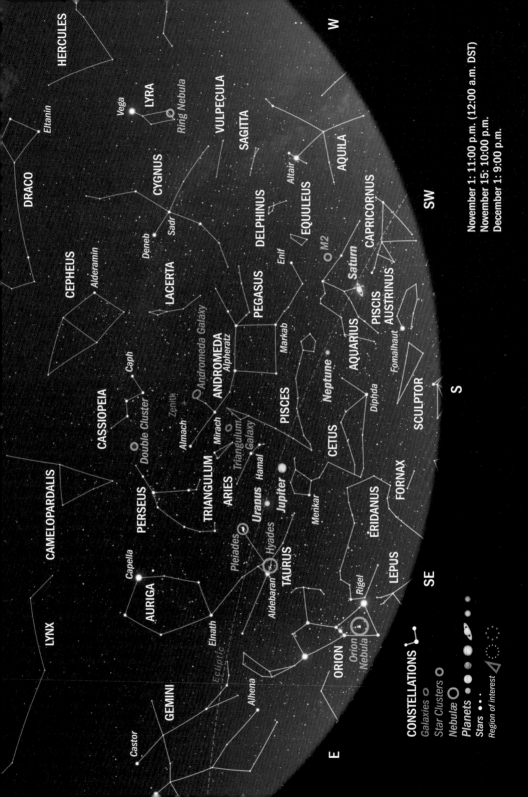

W

HERCULES

DRACO

Eltanin

Vega
LYRA
Ring Nebula

CYGNUS

CEPHEUS

Alderamin

VULPECULA

SAGITTA

Deneb

Sadr

LACERTA

AQUILA

Altair

Caph

Double Cluster

CASSIOPEIA

Andromeda Galaxy

Zenith

ANDROMEDA
Alpheratz

PEGASUS

DELPHINUS

Enif

EQUULEUS

M2

Saturn

PISCIS
AUSTRINUS

CAPRICORNUS

SW

CAMELOPARDALIS

PERSEUS

Almach

Mirach

TRIANGULUM
Triangulum
Galaxy

ARIES
Hamal

Markab

PISCES

CETUS

Diphda

AQUARIUS

Neptune

Fomalhaut

SCULPTOR

S

LYNX

Capella

AURIGA

Elnath

Pleiades

Hyades

Uranus

Jupiter

Menkar

Aldebaran

TAURUS

Menkar

FORNAX

ERIDANUS

LEPUS

SE

GEMINI

Castor

Alhena

Ecliptic

Rigel

ORION
Orion
Nebula

CONSTELLATIONS

Galaxies ○
Star Clusters ○
Nebulæ ○
Planets ● ● ● ● ●
Stars ● ● ● ·
Region of interest

November 1: 11:00 p.m. (12:00 a.m. DST)
November 15: 10:00 p.m.
December 1: 9:00 p.m.

E

The Hyades (Melotte 25) and the Pleiades (Messier 45)

Highlights in the Southern Sky

You can still find **Saturn** sitting snuggly within **Aquarius (the Water Bearer)**. **Jupiter** is also nearby in the faint constellation of **Aries (the Ram)**. **Cetus (the Whale)** is center stage as it lies in the southern sky for most of the early night.

Two fantastic star clusters are now visible, the **Hyades (Melotte 25)** and the **Pleiades (Messier 45)**. The Hyades is an open star cluster found in **Taurus (the Bull)**, and it can easily be found by looking for a bright red star in the constellation named **Aldebaran** (the star itself is not part of the cluster). The Hyades is the closest star cluster to Earth, lying just 153 light-years away. The Pleiades is another close open star cluster, lying 460 light-years from Earth. Photographs of the Pleiades reveal beautiful blue nebulosity around it. Both targets are breathtaking in binoculars.

One of the longest constellations, **Eridanus (the River Eridanus)** winds its way in the south, along the horizon. The constellation begins near the star **Rigel**, the right "foot" of **Orion (the Hunter)**, and winds down to the horizon.

Pegasus, with its distinctive square, is beginning to dip in the southwest.

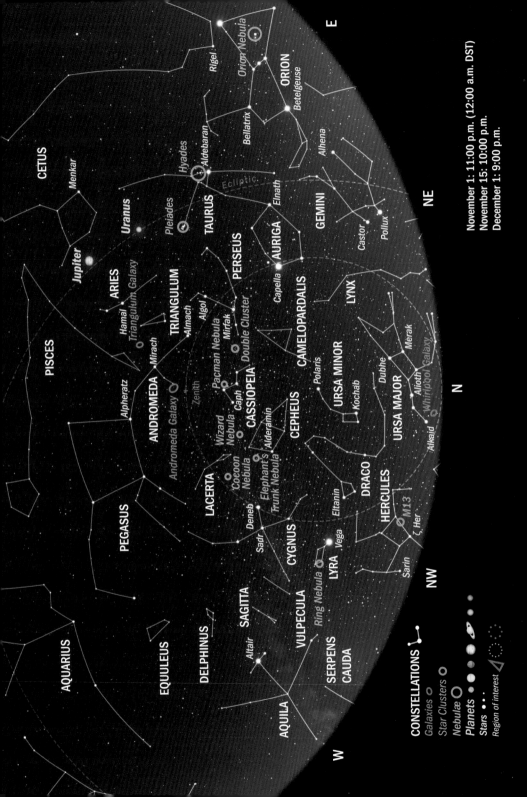

Elephant's Trunk Nebula

Highlights in the Northern Sky

You may notice a bright star in the west. That would be the brightest star of **Lyra (the Lyre)**, known as **Vega**. The constellation is now beginning to disappear from the sky as it sets. **Cygnus (the Swan)** is also following suit, as it gets lower in the northwest.

The little house of **Cepheus** is now in the northwest along with **Cassiopeia**. Using a pair of binoculars, you will find a cornucopia of treasures in the region around these two constellations, with nebulae and star clusters aplenty. Some of the most popular nebula include the **Pacman Nebula (NGC 281)**, the **Wizard Nebula (NGC 7380)**, the **Cocoon Nebula (IC 5146)** and the **Elephant's Trunk Nebula**, which is located in **IC 1396**.

Try to spot the small **Delphinus (the Dolphin)** in the east and **Lacerta (the Lizard)** that lies between Cepheus and **Hercules.**

December

December's Events

Orion returns as king of the night sky this month.

There's plenty to see if you're willing to bundle up and head outside during the colder weather, including many nebulae in the direction of Orion.

Mercury has a greatest elongation (21.3 degrees) east of the Sun on December 4, although it will be very low in the southwest after sunset.

There's also one of the best meteor showers of the year this month: the Geminids. This shower can produce more than 150 meteors per hour. The Geminids are one of the most reliable showers of the year, and you may even spot bright fireballs. The only downside is that this time of year tends to be cloudier in some parts of North America, so there's a gamble. This year the shower peaks on the night of December 14 to 15. The Moon is in crescent phase and the shower radiant rises early in the evening, so there will be dark skies for observing lots of meteors.

Winter solstice is on December 22.

Calendar of Events

Day	Time (UTC)	Event
01	03:23	Pollux 1.7°N of Moon
02	04:21	Beehive 3.9°S of Moon
04	13:59	Mercury 21.3°E of Sun (greatest eastern elongation)
04	18:42	Moon at apogee: 404,300 km (251,220 mi.)
05	05:49	Last quarter of the Moon
08	14:05	Spica 2.5°S of Moon
09	16:53	Venus 3.9°N of Moon
12	23:32	New Moon
15		Geminid meteor shower peak
16	18:53	Moon at perigee: 367,900 km (228,602 mi.)
17	21:58	Saturn 2.5°N of Moon
19	18:39	First quarter of the Moon
22	03:28	Winter solstice
22	14:20	Jupiter 2.7°S of Moon
23		Ursid meteor shower peak
24	07:37	Pleiades 1.1°N of Moon
27	00:33	Full Moon
28	11:51	Pollux 1.9°N of Moon

The Moon This Month

SUN	MON	TUES	WED	THURS	FRI	SAT
					1	2
3	4	5 Last Quarter	6	7	8	9
10	11	12 New Moon	13	14	15	16
17	18	19 1st Quarter	20	21	22	23
24	25	26	27 Full Moon	28	29	30
31						

The Moon has a few close encounters this month. On December 9, the Moon will be roughly 2 degrees from Venus in the morning sky. On December 17, it will be roughly 2.5 degrees away from Saturn just after sunset. And on December 24, it will lie roughly 2 degrees from the Pleiades.

The Moon will be at apogee on December 4. Perigee falls on December 16.

The new Moon is on December 12, and the full Moon is on December 27.

The dark areas of the Moon can be faintly illuminated by earthshine, when the Sun's light reflects off the Earth's surface

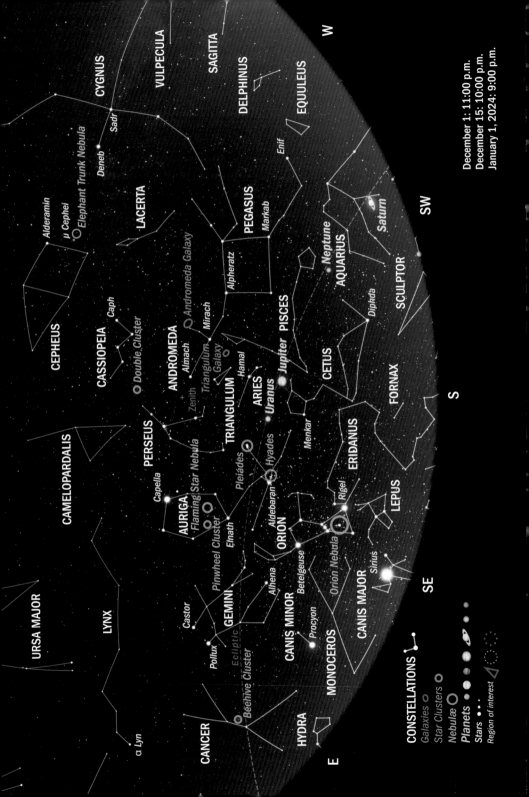

December 1: 11:00 p.m.
December 15: 10:00 p.m.
January 1, 2024: 9:00 p.m.

Flaming Star Nebula (IC 405)

Highlights in the Southern Sky

While all stars may seem to look the same, they're not. Stars come in many different shapes and colors, which you can even see for yourself. The stars in **Orion (the Hunter)** are perfect examples. If you look to the left "shoulder" star of Orion, you'll find **Betelgeuse** (it's safe to say it three times). Betelgeuse is a red supergiant that lies 724 light-years from Earth. Like its summer counterpart Antares (also a red supergiant), its reddish color is clearly apparent. Now, if you look to the right "foot" of Orion, you will see a star that is bright white. That is **Rigel**, a blue-white supergiant that lies 870 light-years from Earth. It is also roughly 47,000 times more luminous than our Sun.

You can also look to **Aldebaran** in **Taurus (the Bull)** for an example of another red star.

Moving away from Taurus, this month **Jupiter** is in the constellation **Aries (the Ram)**.

Auriga (the Charioteer) is also prominent high in the southwest, near Orion. The constellation is easy to spot, especially with its star **Capella,** the sixth-brightest in the night sky. There are some nice targets at which you can point a telescope, should you have one, including the **Flaming Star Nebula (IC 405)**. There's also the **Pinwheel Cluster (Messier 36)**, an open cluster easily seen with binoculars.

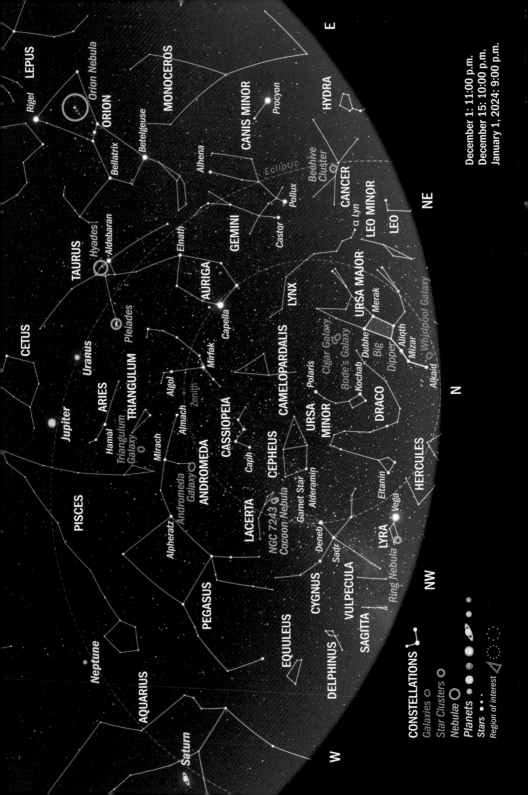

Cocoon Nebula (IC 5146)

Highlights in the Northern Sky

Ursa Major (the Great Bear) along with its asterism the **Big Dipper** is now climbing in the northeast. You may want to see if you can spot **Messier 81 (Bode's Galaxy)** and **Messier 82 (the Cigar Galaxy)** using a telescope. The region actually contains many galaxies, but most are only visible using large telescopes.

Ursa Minor (the Little Bear) looks like a small upside-down spoon now.

To the east is **Gemini (the Twins)** with its bright stars **Castor** and **Pollux**. And joining Gemini is **Cancer (the Crab)**, a faint constellation that is home to the beautiful **Beehive Cluster (Messier 44)**. A pair of binoculars will reveal the many stars in this open cluster.

To the east is **Cepheus**, where you can try to spot its "garnet star," **Mu (μ) Cephei**. Nearby is **Lacerta (the Lizard)**. Use a pair of binoculars to find the open star cluster **NGC 7243**. Just below that is the **Cocoon Nebula (IC 5146)**, a favorite target for many astrophotographers that lies just 2,500 light-years away.

The Messier Catalog

Name	Traditional Name	Type	Constellation
Messier 1 (NGC 1952)	Crab Nebula	Supernova remnant	Taurus
Messier 2 (NGC 7089)		Globular cluster	Aquarius
Messier 3 (NGC 5272)		Globular cluster	Canes Venatici
Messier 4 (NGC 6121)		Globular cluster	Scorpius
Messier 5 (NGC 5904)		Globular cluster	Serpens
Messier 6 (NGC 6405)	Butterfly Cluster	Open cluster	Scorpius
Messier 7 (NGC 6475)	Ptolemy Cluster	Open cluster	Scorpius
Messier 8 (NGC 6523)	Lagoon Nebula	Emission nebula with cluster	Sagittarius
Messier 9 (NGC 6333)		Globular cluster	Ophiuchus
Messier 10 (NGC 6254)		Globular cluster	Ophiuchus
Messier 11 (NGC 6705)	Wild Duck Cluster	Open cluster	Scutum
Messier 12 (NGC 6218)	Gumball Globular	Globular cluster	Ophiuchus
Messier 13 (NGC 6205)	Hercules	Globular cluster	Hercules
Messier 14 (NGC 6402)		Globular cluster	Ophiuchus
Messier 15 (NGC 7078)	Great Pegasus Cluster	Globular cluster	Pegasus
Messier 16 (NGC 6611)	Eagle Nebula	Emission nebula with cluster	Serpens
Messier 17 (NGC 6618)	Omega Nebula	Emission nebula with cluster	Sagittarius
Messier 18 (NGC 6613)		Open cluster	Sagittarius
Messier 19 (NGC 6273)		Globular cluster	Ophiuchus
Messier 20 (NGC 6514)	Trifid Nebula	Emission, reflection and dark nebula with cluster	Sagittarius
Messier 21 (NGC 6531)		Open cluster	Sagittarius
Messier 22 (NGC 6656)	Sagittarius Cluster	Globular cluster	Sagittarius
Messier 23 (NGC 6494)		Open cluster	Sagittarius
Messier 24 (IC 4715)	Sagittarius Star Cloud	Milky Way star cloud	Sagittarius
Messier 25 (IC 4725)		Open cluster	Sagittarius
Messier 26 (NGC 6694)		Open cluster	Scutum
Messier 27 (NGC 6853)	Dumbbell Nebula	Planetary nebula	Vulpecula
Messier 28 (NGC 6626)		Globular cluster	Sagittarius
Messier 29 (NGC 6913)		Open cluster	Cygnus

Name	Traditional Name	Type	Constellation
Messier 30 (NGC 7099)		Globular cluster	Capricornus
Messier 31 (NGC 224)	Andromeda Galaxy	Spiral galaxy	Andromeda
Messier 32 (NGC 221)	Le Gentil	Dwarf elliptical galaxy	Andromeda
Messier 33 (NGC 598)	Triangulum Galaxy	Spiral galaxy	Triangulum
Messier 34 (NGC 1039)		Open cluster	Perseus
Messier 35 (NGC 2168)		Open cluster	Gemini
Messier 36 (NGC 1960)	Pinwheel Cluster	Open cluster	Auriga
Messier 37 (NGC 2099)		Open cluster	Auriga
Messier 38 (NGC 1912)	Starfish Cluster	Open cluster	Auriga
Messier 39 (NGC 7092)		Open cluster	Cygnus
Messier 40	Winnecke 4	Double star	Ursa Major
Messier 41 (NGC 2287)		Open cluster	Canis Major
Messier 42 (NGC 1976)	Orion Nebula	Emission-reflection nebula	Orion
Messier 43 (NGC 1982)	De Mairan's Nebula	Emission-reflection nebula	Orion
Messier 44 (NGC 2632)	Beehive Cluster	Open cluster	Cancer
Messier 45	Pleiades	Open cluster	Taurus
Messier 46 (NGC 2437)		Open cluster	Puppis
Messier 47 (NGC 2422)		Open cluster	Puppis
Messier 48 (NGC 2548)		Open cluster	Hydra
Messier 49 (NGC 4472)		Elliptical galaxy	Virgo
Messier 50 (NGC 2323)	Heart-Shaped Cluster	Open cluster	Monoceros
Messier 51 (NGC 5194, NGC 5195)	Whirlpool Galaxy	Spiral galaxy	Canes Venatici
Messier 52 (NGC 7654)		Open cluster	Cassiopeia
Messier 53 (NGC 5024)		Globular cluster	Coma Berenices
Messier 54 (NGC 6715)		Globular cluster	Sagittarius
Messier 55 (NGC 6809)	Summer Rose Star	Globular cluster	Sagittarius
Messier 56 (NGC 6779)		Globular cluster	Lyra
Messier 57 (NGC 6720)	Ring Nebula	Planetary nebula	Lyra
Messier 58 (NGC 4579)		Barred spiral galaxy	Virgo
Messier 59 (NGC 4621)		Elliptical galaxy	Virgo

Name	Traditional Name	Type	Constellation
Messier 60 (NGC 4649)		Elliptical galaxy	Virgo
Messier 61 (NGC 4303)		Spiral galaxy	Virgo
Messier 62 (NGC 6266)		Globular cluster	Ophiuchus
Messier 63 (NGC 5055)	Sunflower Galaxy	Spiral galaxy	Canes Venatici
Messier 64 (NGC 4826)	Black Eye Galaxy	Spiral galaxy	Coma Berenices
Messier 65 (NGC 3623)		Barred spiral galaxy	Leo
Messier 66 (NGC 3627)		Barred spiral galaxy	Leo
Messier 67 (NGC 2682)	King Cobra Cluster	Open cluster	Cancer
Messier 68 (NGC 4590)		Globular cluster	Hydra
Messier 69 (NGC 6637)		Globular cluster	Sagittarius
Messier 70 (NGC 6681)		Globular cluster	Sagittarius
Messier 71 (NGC 6838)		Globular cluster	Sagittarius
Messier 72 (NGC 6981)		Globular cluster	Aquarius
Messier 73 (NGC 6994)		Asterism	Aquarius
Messier 74 (NGC 628)	Phantom Galaxy	Spiral galaxy	Pisces
Messier 75 (NGC 6864)		Globular cluster	Sagittarius
Messier 76 (NGC 650, NGC 651)	Little Dumbbell Nebula	Planetary nebula	Perseus
Messier 77 (NGC 1068)	Cetus A	Spiral galaxy	Cetus
Messier 78 (NGC 2068)		Reflection nebula	Orion
Messier 79 (NGC 1904)		Globular cluster	Lepus
Messier 80 (NGC 6093)		Globular cluster	Scorpius
Messier 81 (NGC 3031)	Bode's Galaxy	Spiral galaxy	Ursa Major
Messier 82 (NGC 3034)	Cigar Galaxy	Starburst irregular galaxy	Ursa Major
Messier 83 (NGC 5236)	Southern Pinwheel Galaxy	Barred spiral galaxy	Hydra
Messier 84 (NGC 4374)		Lenticular or elliptical galaxy	Virgo
Messier 85 (NGC 4382)		Lenticular or elliptical galaxy	Coma Berenices
Messier 86 (NGC 4406)		Lenticular or elliptical galaxy	Virgo
Messier 87 (NGC 4486)	Virgo A	Elliptical galaxy	Virgo

Name	Traditional Name	Type	Constellation
Messier 88 (NGC 4501)		Spiral galaxy	Coma Berenices
Messier 89 (NGC 4552)		Elliptical galaxy	Virgo
Messier 90 (NGC 4569)		Spiral galaxy	Virgo
Messier 91 (NGC 4548)		Barred spiral galaxy	Coma Berenices
Messier 92 (NGC 6341)		Globular cluster	Hercules
Messier 93 (NGC 2447)		Open cluster	Puppis
Messier 94 (NGC 4736)	Cat's Eye Galaxy	Spiral galaxy	Canes Venatici
Messier 95 (NGC 3351)		Barred spiral galaxy	Leo
Messier 96 (NGC 3368)		Spiral galaxy	Leo
Messier 97 (NGC 3587)	Owl Nebula	Planetary nebula	Ursa Major
Messier 98 (NGC 4192)		Spiral galaxy	Coma Berenices
Messier 99 (NGC 4254)	Coma Pinwheel	Spiral galaxy	Coma Berenices
Messier 100 (NGC 4321)		Spiral galaxy	Coma Berenices
Messier 101 (NGC 5457)	Pinwheel Galaxy	Spiral galaxy	Ursa Major
Messier 102 (NGC 5866)	Spindle Galaxy	Lenticular galaxy	Draco
Messier 103 (NGC 581)		Open cluster	Cassiopeia
Messier 104 (NGC 4594)	Sombrero Galaxy	Spiral galaxy	Virgo
Messier 105 (NGC 3379)		Elliptical galaxy	Leo
Messier 106 (NGC 4258)		Spiral galaxy	Canes Venatici
Messier 107 (NGC 6171)		Globular cluster	Ophiuchus
Messier 108 (NGC 3556)	Surfboard Galaxy	Barred spiral galaxy	Ursa Major
Messier 109 (NGC 3992)		Barred spiral galaxy	Ursa Major
Messier 110 (NGC 205)	Edward Young Star	Dwarf elliptical galaxy	Andromeda

Glossary

annular solar eclipse: A kind of partial solar eclipse during which the Moon does not completely cover the Sun's disk but leaves a ring of sunlight around the Moon.

aphelion: For an object orbiting the Sun, the point in its orbit that is farthest from the Sun.

apogee: When the Moon (or any satellite of Earth) is at its most distant in its monthly orbit around Earth.

arcsecond (or second of arc): A tiny angle equal to 1/3600 of a degree, used to measure the separation of double stars and the apparent diameters of solar system objects.

asterism: a group of stars within a constellation (or sometimes from more than one constellation) that forms its own distinct pattern.

astronomical unit (AU): A unit of distance that uses the average distance between Earth and the Sun as its base metric. One astronomical unit is approximately 150 million kilometers (93.2 million miles).

autumnal (fall) equinox: The date in September marking the end of summer and the beginning of autumn, when the Sun illuminates the Northern and Southern Hemispheres equally. The lengths of the day and night are equal, and it's one of two instances of the year that the Sun has a declination of 0 degrees.

binary star system: A double-star system in which two stars are gravitationally bound to each other and orbit a common center of mass. Many double-star systems have multiple components.

conjunction: Technically speaking a conjunction is when two objects have the same right ascension, but amateur astronomers also use the term when there is a close approach of two or more celestial objects.

constellation: A group of stars that make an imaginary image in the night sky. The International Astronomical Union (IAU) has designated the boundaries between 88 official constellations covering the celestial sphere.

degree: A unit of measurement used to measure the distance between objects or the position of objects in astronomy. The entire sky spans 360 degrees, and up to about 180 degrees of sky is visible from any given point on Earth with an unobstructed horizon.

double star (binary star): A pair of stars with small angular separation (from fractions of an arcsecond to hundreds of arcseconds). Optical doubles are chance alignments of stars with no physical connection to each other. True binary stars are gravitationally bound (see binary star system).

fireball: A very bright meteor, generally brighter than magnitude –4.0 (about the brightness of Venus).

galaxy: An enormous system of gas, dust and billions of stars and their planetary systems, all held together by gravity.

globular cluster: Old star systems at the edges of spiral galaxies that can contain anywhere from thousands to millions of stars, packed in a close, roughly spherical form and held together by gravity.

greatest eastern elongation: When an inner planet (Mercury or Venus) is farthest from the Sun in the western evening sky.

greatest western elongation: When an inner planet (Mercury or Venus) is farthest from the Sun in the eastern morning sky.

light-year: A unit of distance that uses the distance light travels in one Earth year as its base metric. One light-year is about 9 trillion kilometers (6 trillion miles). Objects in outer space are so far apart that it takes a long time for their light to reach Earth, and so the farther an object is, the farther in the past that observers on Earth are seeing it. As an example, the Andromeda Galaxy (Messier 31) is 2.5 million light-years away, so when observers look at it in the sky, they are seeing it as it appeared 2.5 million years ago.

magnitude: The apparent brightness of an object in the sky. Magnitude is measured on a scale where the higher the number, the fainter the object appears. Objects with negative numbers are brighter than those with positive numbers.

meteor shower: An atmospheric event during which a number of meteors can be seen radiating from one point in the sky. Meteor showers occur when Earth, at a particular point in its orbit, crosses a stream of particles left over from a passing comet or asteroid.